AS Sociology
UNIT 3

Module 3: Sociological Methods

Joan Garrod

Philip Allan Updates
Market Place
Deddington
Oxfordshire
OX15 0SE

tel: 01869 338652
fax: 01869 337590
e-mail: sales@philipallan.co.uk
www.philipallan.co.uk

This Guide has been written specifically to support students preparing for the AQA AS Sociology Unit 3 examination. The content has been neither approved nor endorsed by AQA and remains the sole responsibility of the author.

Typeset by Magnet Harlequin, Oxford
Printed by Raithby, Lawrence & Co. Ltd, Leicester

P00026

Contents

Introduction

■ ■ ■

Content Guidance

■ ■ ■

Questions and Answers

■ ■ ■

AS Coursework Task

Introduction

About this guide

This unit guide is for students following the AQA AS Sociology course. It deals with the Module 3 topic **Sociological Methods**, which is examined either by a written examination or by the submission of a structured coursework task. There are four sections to this guide:

- **Introduction** — this provides advice on how to use this unit guide, an explanation of the skills required in AS Sociology and suggestions for effective revision for the written examination, or preparing for the coursework task. It concludes with guidance on how to do well in the unit test.
- **Content Guidance** — this provides an outline of what is included in the specification for Sociological Methods. It is designed to make you aware of what you should know before the unit test or before writing up your coursework task.
- **Questions and Answers** — this provides exam-type questions on Sociological Methods for you to try, together with sample answers at grade-A and grade-C level. Examiner's comments are included to show how the marks have been awarded.
- **AS Coursework Task** — this provides an example of a candidate's first rough draft, with some hints and advice and some exercises for you to try.

How to use the guide

To use this guide to your best advantage, from the beginning of your study of Sociological Methods you should refer to the Introduction and Content Guidance sections. The Content Guidance section gives you more detail about the bullet points in the specification — in other words, it tells you what you need to know to be certain that you have covered the necessary material. Remember that exam questions can be set on any aspect of the specification, so you must ensure that you have covered all of the bullet points for each topic area you are studying. The Content Guidance section also gives you some important concepts associated with the topic of Sociological Methods, but it is important to remember that these are meant only as a guide — you will almost certainly come across others.

In order to get full advantage from the Question and Answer section, you would be well advised to wait until you have completed your study of the topic, as the questions are wide ranging and will cover more than one aspect of the specification for Sociological Methods. When you are ready to use this section, you should take each question in turn, study it carefully and either write a full answer yourself (which is the best option) or, at the very least, answer parts (a) to (d) fully and write a plan for parts (e) and (f). When you have done this, study the grade-A candidate's answer and compare it with your own, also paying close attention to the examiner's comments. You could then look at the grade-C candidate's answer and, using the examiner's comments as a guide, work out how to rewrite the answer so that it would gain higher marks.

These tasks are quite intensive and time consuming, and you should not be tempted to try to tackle all the questions in a short space of time. It is better to focus on one at a time, and spread the workload over several weeks — you can always find some time to do this, even while studying another topic. In addition to using the questions to consolidate your own knowledge and develop your exam skills, you should use at least some of the questions as revision practice — even just reading through the grade-A candidates' answers should provide you with useful revision material.

If you are going to undertake the coursework task, rather than sit the written exam, then you should study the final section of the book before you embark on developing your own proposal. In particular, you should familiarise yourself with the mark scheme for each of the four coursework sections, and pay careful attention to the maximum word length for each. Your teacher will advise you on how the coursework task is to be managed in your own institution, but you should always discuss your aim/hypothesis and your first rough draft with your teacher before you attempt to write up your proposal, and of course follow any advice you are given.

The AS specification

The AS module on Sociological Methods is designed to give you a good understanding of the range of methods employed by sociologists in their research, the strengths and weaknesses of different methods, and the reasons influencing the choice of method. You should be able to illustrate your discussion of any particular method by referring to at least one study in which the method has been employed. As with all modules in the specification, as well as displaying appropriate knowledge and understanding in the examination or in your coursework task, you will need to display a range of skills, as outlined below.

Examinable skills

There are two main examinable skills in the specification, also referred to as 'Assessment Objectives'. Each of these counts for half of the available marks, both within a question and in the AS qualification as a whole.

Assessment Objective 1 (AO1) is 'knowledge and understanding'. This skill is not quite as straightforward as it seems. You not only need to have sufficient 'knowledge' of the topic, you also need to show that you actually understand it and are not just putting something down that you have learned by rote. Understanding, then, is shown by the way in which you select and use your knowledge to answer a particular question or to complete your coursework task. It must be relevant knowledge: that is, appropriate to the question you are answering or the aim or hypothesis that you have chosen. You must also select and use appropriate concepts, to show that you have studied and understood a topic in a sociological — that is, a specialist — way.

Assessment Objective 2 (AO2) covers the skills of identification, analysis, interpretation and evaluation. In terms of how you display these skills in the exam or in your coursework task, you must learn to:

- **identify** appropriate pieces of knowledge
- **distinguish** between facts and opinions
- **analyse** research methods and research studies in terms of their strengths and weaknesses
- **interpret** material such as research findings and statistics in order to identify any trends and uncover the (sometimes hidden) meanings
- **evaluate** all the material you come across during your course of study

The skill of evaluation is a very important one. Right from the start, you should develop the habit of asking questions, such as 'Who says so?', 'How did they find that out?', 'Is there any other evidence of this?', 'What do other sociologists think?' and so on. In perhaps more practical terms, it means that whenever you are introduced to a sociological method, perspective or study, you should find and learn at least two criticisms that have been made of it. You should also note, of course, which group or person has made these criticisms, as this is an important piece of information.

Revision planning

Good preparation for revision actually starts the minute you begin to study sociology. One of the most important revision aids that you will have is your sociology folder, so it is important that you keep this in good order. Essentially, it should be divided into topic areas. It should contain all your class notes, handouts, notes you have made from textbooks, class and homework exercises and, of course, all your marked and returned work. If you are not by nature a neat and tidy person, you may find that you have to rewrite notes you made in class into a legible and coherent form before putting them in your folder. Be warned, though — this is something you should do straight away, as even after only a few days you will have forgotten things. If you keep a good folder throughout, reading through this will form a major part of your revision. In addition, you will, of course, need to re-read the relevant parts of your textbooks. Your own work also forms an important revision resource. Go back over your essays and exam answers, read your teacher's comments and use them to see whether you can redo any pieces that did not score particularly good marks.

Another important aspect of revision is to practise writing answers within the appropriate time limit. Make sure you have sufficient time not only to complete all the parts of the question, but also to re-read your answer, in order to correct any mistakes that may have crept in while working under pressure.

Finally, you need to ensure that you have a thorough understanding of a range of appropriate concepts and studies. Again, this planned and comprehensive revision is not something that can be done the night before the exam — you should start at least a couple of weeks before the exam, and revise in concentrated bursts of time. People differ in this respect, but it is seldom a good idea to spend more than 2 hours at a time on revision. For most people, two or three stints of an hour at a time spread out over a day or two will be more productive than a 2- or 3-hour session, particularly late at night.

The unit test

As the only topic in Module 3 is Sociological Methods, and the choice is between taking the unit test or submitting the coursework task, the unit test contains only one question, to be completed in 1 hour. The question is marked out of 60, and of these 60 marks, 30 are given to AO1 (knowledge and understanding) and 30 to AO2 (identification, analysis, interpretation and evaluation). The unit is worth 30% of the AS qualification, and 15% of the total A-level qualification.

The question will contain source material, or 'items' — usually two of them. These are designed to help you by providing information on which you should draw in your answer. It is therefore essential that before attempting to answer any part of the question, you read the items carefully, and continue to refer back to them throughout the examination. Sometimes a question will make a specific reference to an item, such as 'With reference to Item A', or 'Using material from Item B and elsewhere'. In these cases you should make quite sure that you have followed the instruction. An easy way of doing this is to say, for example, 'Item A shows evidence of...', or 'Item B provides an example of how participant observation was used to...'.

The source material in the Unit 3 test will often summarise or describe an example of sociological research using a particular method, or give details of a 'research context': that is, a situation which a sociologist wishes to research or on which research has already been done. The subsequent questions will often link closely to this material, so that you may be required, for example, to summarise the advantages of using a particular method 'in the research context outlined in the item', or to suggest problems likely to have been encountered by the authors of the research described in the item.

The question is broken down into a number of parts, usually (a) to (f), each with its own mark allocation. The first series of questions, typically (a) to (d), will together add up to 20 marks, and require short answers. As is usual in such cases, the greater the number of marks allocated, the more you will normally need to write to gain full marks. Beware of writing more than is necessary, particularly for questions (a), (b) and (c). You should normally write two short paragraphs for question (d).

The remaining parts, (e) and (f), each carry 20 marks, and there is an important differ-ence between them with regard to the balance of the two assessment objectives.

Part (e) is weighted towards knowledge and understanding, which carries 14 of the 20 available marks, leaving 6 marks for the AO2 skills. This means that in answering this question, you should ensure that you show evidence of appropriate sociological knowledge and understanding of research methods. Remember, though, that 6 marks are for demonstration of the remaining skills, so these, particularly evaluation, should not be neglected. You should therefore make sure that you include some critical or evaluative comments at suitable points in your answer.

In part (f), the skills balance is reversed: that is, 14 of the 20 marks are awarded for the skills of identification, analysis, interpretation and evaluation, with 6 marks avail-able for knowledge and understanding. As it is most unlikely that you will be able to demonstrate the AO2 skills without at the same time showing evidence of knowledge and understanding (though this must, of course, be relevant to the question), your focus should be on showing evidence of the range of AO2 skills. You will be reminded of the importance of this by the wording of the question, which will usually ask you to 'assess' or 'evaluate' something.

Content Guidance

This section will take you through the bullet points in the AQA specification for the topic of Sociological Methods. It will indicate what you should know about the different methods and will list some important concepts. Remember, though, that there are other concepts which are useful, and you should note these from your textbooks and handouts and make sure that you understand them. You should also make sure that you have a section in your folder for 'Methods', and that you note down the methods used in the various studies you look at while working on other parts of the course. If you have back copies of *Sociology Review* or *S Magazine* in your library, you will find some useful articles about different methods, and if you are a current subscriber to *Sociology Review*, you will continue to find helpful material.

Sociological research

Forms of data

Sociological research can be based on primary or secondary data. **Primary data** refers to information that the researcher has generated by his or her own research, such as the results of a survey or a series of interviews, experiments or observations. **Secondary data** refers to information that already exists and that the researcher has used. Commonly used sources of secondary data are official statistics and reports, the content of newspapers and magazines, personal documents such as diaries and letters, historical documents and research studies by other sociologists.

Sociological research can produce either quantitative or qualitative data.

- **Quantitative data** are those that provide numerical information. This can be displayed in various ways, such as tables, graphs, bar charts, pie charts, tally charts, columns of figures and lists of percentages. Common sources of quantitative data in sociology are official statistics and the results of surveys or structured interviews using closed questions.
- **Qualitative data** consist of information in the form of a description in words or images. The most common forms of qualitative data are transcripts or summaries of interviews or conversations, the researcher's description of a place, a group or a situation, and a 'reading' of a text or image.

Reliability and validity

Reliability and validity are important terms that can be applied both to sociological methods and to the data that they generate.

- **Reliability** refers to the degree to which something, if repeated, would give the same or very similar results. It is usually assumed that, given a carefully chosen and representative sample and a well-constructed questionnaire, large-scale social surveys are reliable.
- **Validity** refers to the extent to which research measures what it actually sets out to measure, and the extent to which the findings are a true reflection of people's beliefs, attitudes and behaviour. It is usually assumed, again given carefully planned and executed research, that qualitative methods, such as unstructured interviews and participant observation, produce valid data.

Remember that something can be reliable without being valid. For example, a questionnaire on domestic violence might, if repeated, give the same results, but the findings would probably not be valid, as questionnaires are not an appropriate way of finding out about the nature and extent of domestic violence.

Key concepts

Primary data; secondary data; quantitative data; qualitative data; reliability; validity.

Research methods based on primary data

Social surveys

Social surveys may be large scale or small scale in nature, and can generate either quantitative or qualitative data, or a mixture of both.

Survey populations

A survey population contains every one of the individuals or groups in which the sociologist is interested. So, a sociologist interested in carrying out a survey on the effects of divorce on children would have a survey population of all children whose parents had been divorced.

Sampling

In most research, it is not possible or practical to carry out research on every unit of the population in which a sociologist is interested, as the example of the children of divorce indicates. Sociologists therefore take a sample, or selection, of the whole population, and conduct research only on that sample. To draw most types of sample requires a **sampling frame**: that is, a list of all the units in the survey population.

In most cases, particularly large-scale surveys, an attempt is made to make the sample **representative** of the whole population, which means that the results of the research into the sample can be applied to the population as a whole. Occasionally, a sociologist is more interested in being able to do research on a particular group, such as a group of travellers or asylum seekers, than in ensuring that the group is representative of all such groups. Sometimes it is not possible to take a representative sample because of the lack of a suitable or available sampling frame, such as in the case of people who are HIV positive, or the victims of domestic violence.

Sampling methods
- **Random (probability) sampling** — this is one of the simplest and most commonly used forms of sampling. Every unit in the survey population is given a number, and a sample of predetermined size is drawn, usually by using tables of random numbers. In this type of sampling, every unit in the survey population has a known and equal probability of being selected. This does not necessarily produce representative samples, unless the sample size is large in proportion to the number in the population.
- **Systematic sampling** — this is useful where the sample population is in the form of a list, such as all the students in a college being listed in a series of class registers. If, for example, the sample was going to be 1 in 20 of the survey

population, a number would be chosen at random between 1 and 20 — say, 14. Number 14 in the first register would be the first sampling unit, then 14 + 20, i.e. number 34, then 34 + 20, i.e. 54 (by which time you would probably be on to the second register) and so on, until the sample size was reached.

- **Stratified random sampling** — this method is possible where the survey population can be divided into mutually exclusive groups, such as by age-bands, or sex, or place of birth. A random sample is then taken from each group in the proportion in which that group appears in the survey population. Hence this type of sampling is highly representative.
- **Multi-stage sampling** — this is when an initial sample is taken, then that sample is in turn sampled to provide a smaller group and so on.
- **Cluster sampling** — this is sometimes used when the survey population is particularly large. The survey population is first divided into smaller units, or clusters, and then a random sample is taken from each.
- **Snowball sampling** — also sometimes called opportunity sampling. This is usually used in cases where no sampling frame exists. It starts with a single contact person, who then puts the researcher in contact with another suitable person, who then provides further contacts and so on. It is a useful method in certain circumstances, though there is no way of knowing the extent to which the sample is representative.
- **Self-selecting sampling** — this is where people respond to an advertisement requesting individuals who have certain characteristics or experiences to volunteer for the research. Again, it is unlikely to be representative.
- **Quota sampling** — this is used more in market research than sociology, and involves the researcher choosing in advance certain characteristics (age, sex, occupation, ethnic group, marital status, etc.) and determining how many people with each characteristic should be surveyed. Interviewers are then usually sent out into the streets to accost passers-by who appear to have the desired characteristics. Each interviewer has to fill his or her 'quota'.

Key concepts

Survey population; sampling frame; sample; representative sample.

Questionnaires

Questionnaires are frequently used in social surveys, particularly where the sample size is large. It is important to remember that a questionnaire is a tool, not a research method in its own right. Questionnaires can consist of **closed questions**, in which the respondents may only choose answers from those provided, or **open-ended questions**, which leave the respondents free to answer in their own words, or, as is often the case, a mixture of both. Closed questions will generate quantitative data, while open-ended questions generate qualitative data. Constructing a good questionnaire is not easy, and most surveys will carry out a **pilot study** to try out the questionnaire on a smaller group before it is used in the main survey.

Questionnaires may be administered by the researchers or by people especially trained for the task, or they may be sent to people through the post. **Postal** (or mailed) question-naires are cheaper than face-to-face questionnaires and can easily cover a wide geographical area, but they typically have a low **response rate** (the proportion of the sample that actually takes part in the survey), which can affect the reliability of the results.

Key concepts

Open-ended questions; closed questions; pilot study; response rate; reliability.

Evaluation

+ Questionnaires can cover a large number of people.
+ They can gather a large amount of information relatively quickly.
+ If the sample is carefully chosen and sufficiently large, a questionnaire is representative.
+ They can help to reduce interviewer bias.
+ Postal questionnaires are relatively cheap.
− Closed questions may not reflect respondents' real views.
− Reliability may be achieved at the expense of validity.
− Bias may be present through the choice of questions asked, the way the questions are phrased, and respondents' reactions (possibly unconscious) to the social characteristics of the interviewer.
− Postal questionnaires may be completed by someone other than the person selected.
− Postal questionnaires tend to have a low response rate.

Interviews

These can be structured or unstructured, or sometimes a mixture of both, in which case the interviews are often referred to as semi-structured. While some research uses group interviews, most interviews are conducted with just one person or, in some family research, a couple.

- **Structured interviews** are very similar to the administering of a questionnaire, in that the researcher has a list of questions that are put in the same order and using the same words to all the respondents. Such a list of questions is known as an interview schedule.
- **Unstructured interviews** resemble much more a conversation between the researcher and the interviewee. The researcher will have a general notion of the topics to be covered, and may give gentle 'prompts', or ask an open-ended question, but usually the point of an unstructured interview is that the interviewee is free to raise any issue. Unstructured interviews are usually recorded.
- **Semi-structured interviews**, as the name implies, are likely to contain both a list of questions and an opportunity for the interviewee to have a free discussion.

While structured interviews may be more reliable, unstructured interviews are considered to have greater validity.

Key concepts

Structured interviews; unstructured interviews; interview schedule; interviewer bias.

Evaluation

+ As interviews are usually conducted in the interviewee's home, people are likely to be more relaxed and feel less pressured. This is particularly true if the interview is being recorded, rather than the interviewer having to write everything down.
+ As interviews usually take longer than the administration of a questionnaire, more issues can be explored, or issues can be explored in more depth.
+ Interviewers can take clues from people's tone of voice and body language as well as from what they say.
+ Unstructured interviews may throw up issues worthy of exploration of which the researcher had not thought.
− The length of time involved and the cost of interviews usually mean that the research findings are based on a relatively small number of respondents.
− Interviews, particularly unstructured ones, generate a huge amount of data, which is very time consuming to analyse.
− There is more chance of interviewer bias, particularly in the case of unstructured interviews.
− Researchers have to select what they consider significant from all the data they receive, and different researchers might consider different things to be significant.
− It is difficult for anyone else to check the researcher's analysis and conclusions.

Non-participant observation

As the name implies, non-participant observation consists of the researcher looking at particular social contexts and situations, and subjecting what is seen to sociological analysis. Observation can produce quantitative data, such as when the researcher is using a tally chart to record the number of incidences of a particular type of behaviour or interaction, and/or qualitative data, when what is seen is described and analysed.

Key concepts

Tally chart; significant behaviour; Hawthorne effect.

Evaluation

+ It is a simple and cheap method to use.
+ If people are unaware that they are being observed, they will act as they usually do.
+ It can be used to analyse social interaction in a wide variety of contexts.
− It relies heavily on the researcher's interpretation of what is going on — there is seldom corroboratory evidence.
− The researcher selects what actions are deemed to be significant.
− The researcher may interpret/analyse the interactions from a particular gender/ethnic/class viewpoint, which will affect the validity of the research.
− If people know that they are being observed, they may act differently from usual.

Participant observation

This may be overt or covert.

- **Overt participant observation** refers to situations in which those observed know that they are being observed.
- **Covert participant observation** refers to situations in which the role of the researcher is hidden from those being studied.

As the name implies, participant observation requires the researcher to take part in at least some of the activities of the group being studied. The researcher must therefore find a suitable role to play. In overt participant observation, this can be the 'true' role, i.e. that of researcher, or it can be a misrepresentation. This can be a small misrepresentation, such as claiming to be someone 'writing a book' without specifying that the book will be wholly about the group, or it can be greater, such as misleading the members of the group about exactly what it is that the researcher is interested in. Overt participant observation requires the researcher to gain the trust and acceptance of the group, while covert participant observation requires the researcher to 'get in, stay in and get out', all of which can be quite difficult. Participant observation is generally regarded as having a high degree of validity.

Key concepts

Overt participant observation; covert participant observation; role; empathy; Hawthorne effect; subjectivity; objectivity; ethics.

Evaluation

Overt participant observation

+ It allows the researcher to see and study at first hand the interactions between group members over a period of time.
+ As the subjects get more used to the presence of the researcher, they are more likely to act naturally.
+ As the people involved know that they are the subject of research, the researcher is able to ask questions.
+ If the subjects are told the truth about the research, it is an ethical research method.
− As the subjects know that they are being studied, they might modify their behaviour (Hawthorne effect), and the researcher would be unaware of this.
− The researcher asking questions might make the subjects more reflective about their own attitudes and actions, which could cause these to change.
− Some groups have the power to refuse access to researchers, or to grant access only under their conditions. This leads to the majority of studies being carried out on the relatively powerless.
− It is usually very time consuming.

Covert participant observation

+ As the subjects are unaware of the presence of a researcher, they will speak and behave as normal.

+ The researcher is likely to learn things that no other method of research would reveal.
+ The method allows research into groups that might be impossible to study by any other means.
− It is unethical, as the subjects have been unable to give their 'informed consent' to the research.
− As the research has to be kept secret from the subjects, it is difficult to find ways of writing down what has been observed.
− The researcher may come to over-empathise with the group, and lose objectivity.
− The fact that the researcher has to play a role can itself alter the dynamics of the group.
− The researcher might be forced to take part in dangerous or criminal activities rather than 'blow the cover'.
− The researcher's interpretation of what is going on within the group could be mistaken.
− It is usually very time consuming.

Experiments

The experimental method is used extensively in the natural sciences, but it is not used widely in sociology. There are two types of experiment: laboratory experiments and field experiments.

- **Laboratory experiments** are carried out under controlled conditions, in contexts where the researcher is able to control all, or most, of the variables. Most laboratory experiments are conducted to test a hypothesis or to monitor the effects of something.
- **Field experiments** are carried out in the 'real world', in natural conditions such as a classroom, where the researchers have been able to set up a situation that creates an experimental group and a control group. The 1968 study by Rosenthal and Jacobson, *Pygmalion in the Classroom*, is an example of a field experiment.

Key concepts

Laboratory experiments; field experiments; scientific method; hypothesis; laws; experimental group; control group; variables; ethics.

Evaluation

+ If the experiment has been carried out under rigorous conditions, it could prove (or disprove) a particular hypothesis and/or establish cause and effect.
+ Field experiments can provide useful information about certain processes, e.g. the effects of teacher expectations.
− It is not normally possible to study humans in a laboratory situation without it modifying their behaviour to an unknown degree. The laboratory is not a 'natural' environment for social interaction.

- If the subjects are unaware that they are taking part in an experiment, it is unethical.
- Some experimental situations can have significant and detrimental effects on some of the participants (e.g. the 1973 'prison experiments' of Haney, Banks and Zimbardo). This again makes the research unethical.
- It is quite impossible for researchers to control, or even be aware of, all the variables influencing the subjects' behaviour, which casts doubt on any conclusions drawn.

The comparative method

The comparative method can be based on primary or secondary data, or a combination of both. It is similar in some ways to the field experiment, with the important difference that the researcher uses a 'real', naturally occurring situation, rather than an artificial one. This method is used to compare one social situation or event with another. Sometimes it is used to make comparisons between two or more things occurring or existing at the same point in time. For example, the comparative method might be used to compare classroom interaction, or exam results, or other factors occurring in mixed (co-educational) and single-sex schools in the same area. Comparisons might also be made on a societal level: for example, looking at family structures, or divorce rates, or crime rates in different societies. The method is also used to compare things over time, in a 'before and after' study. For example, Willmott and Young looked at the effects on family structure and family life when people moved out of inner-city Bethnal Green into the suburbs of Greenleigh.

Key concepts

Causal relationships; spurious correlation.

Evaluation

+ It can help to identify some of the causes of social phenomena (but see below).
+ It makes use of naturally occurring situations rather than artificially constructed ones.
- So many variables are involved in social situations that it is often very difficult to identify the 'cause' of something.
- If the comparison is made using secondary data, particularly statistics, it is open to all the possible weaknesses of this type of information.

Longitudinal studies

This refers not to a particular method as such, but to any study that is carried out on a particular group over a period of time. It is thus quite different from most research which provides a 'snapshot' of a single point in time. Longitudinal studies often involve a variety of different methods, such as questionnaires, interviews and subject diaries. The significant feature is the length of time over which the study is conducted. The National Child Development Study, for example, is still conducting research on the original members of the survey, who were born in 1958. As these studies are difficult to set up and very costly to operate, they mostly have some kind of national or government funding. The Census is an example of a longitudinal study, giving a picture of different aspects of our society at 10-year intervals.

Key concepts

Time-scale; correlation; variables.

Evaluation

+ The length of time involved makes it possible to 'track' what happens to individuals over time and, if the sample is sufficiently large, to begin to identify certain variables that appear to have a significant effect on people's lives.
− The method is extremely costly.
− It is difficult to keep all the original members of the group involved in the study, which can affect the representativeness of the sample.
− The work is usually carried out by teams of researchers, with changes of personnel over time, possibly leading to changes in practice that make comparisons with earlier data more difficult.
− Changes in the funding providers (e.g. changes in government) might lead to different priorities being imposed, again making comparisons more difficult, or even taking the research into new areas.

Research methods based on secondary data

Content analysis

This method is particularly associated with research into the mass media, but it can be used with a wide variety of documents and research topics. Content analysis can generate both quantitative and qualitative data.

Quantitative data result from the calculation of the quantity in given units, or the number of times certain predetermined factors occur in the material being analysed. For example, the Glasgow Media Group measured the amount of time given in television news to certain issues associated with the miners' strike of the early 1980s. Sociologists interested in crime have measured, in column inches, the amount of space devoted in different newspapers to particular types of crime.

Qualitative data result from the researcher's 'reading', or interpretation, of the 'message' or ideological content of the material. For example, the Glasgow University Media Group showed how certain viewpoints and interpretations of events were prioritised over others, and how interviews with miners' representatives, usually conducted outdoors on noisy picket lines, contrasted with interviews with representatives of the employers, which took place indoors in 'business' settings. From this, they attempted to show how such images formed different impressions in viewers' minds about the kind of people on both sides of the dispute. Ferguson analysed the content of women's magazines over 30 years, and showed how the (assumed) focus of 'women's interests' had changed in that time.

Key concepts

Mass media; audience; text; message; ideology; semiology; bias; coding.

Evaluation

Content analysis with quantitative data

+ It is a relatively cheap and simple method to use.
+ It provides statistical evidence of features that may not be readily apparent.
+ It enables comparisons to be made over time.
− The researcher has to decide on the coding categories to be used, and bias may occur.
− It is often difficult to decide precisely into which category something falls, and if more than one researcher is involved, they may not make the same type of decision.
− It is 'on the surface' and does not examine how the audience receives and interprets the material.

Content analysis with qualitative data

+ It allows the 'unpacking' of material to show hidden messages and biases.
+ It goes beyond a surface reading of the text.
− It is the researcher's own interpretation and analysis; others might see things differently.
− Follow-up research is not always carried out to see exactly what 'message' the audience is receiving.

Statistics

Much sociological research based on statistics uses official statistics: that is, statistics produced and published by the government or government agencies. It is important to remember, however, that statistics produced by other bodies, such as charities, trade unions and pressure groups, are also used. Statistics are often classified as either 'hard' or 'soft'.

- **'Hard' statistics** are those that relate to relatively clear-cut events, such as rates of birth, death, marriage, divorce, passes in public examinations and so on.

- **'Soft' statistics** are those for which there is a greater degree of subjectivity in their collection and presentation. Examples are statistics relating to crime, poverty, welfare benefits, unemployment and health.

Interpretivists have pointed to the fact that statistics, particularly 'soft' statistics, are social constructs, and cannot, therefore, be treated as 'scientific' data.

Key concepts

Official statistics; hard statistics; soft statistics; rate; objectivity; subjectivity; social construct.

Evaluation

- + Statistics are easy to access and are either free or relatively inexpensive.
- + Researchers can gain access to data they would not be able to find out for themselves.
- + Comparisons can be made over time and within and between countries.
- + Positivists regard official statistics in particular as an objective, 'scientific' data source.
- − The classifications of different events can, and do, change over time, making real comparisons impossible.
- − Where statistics are gathered by a number of people, they may not be using exactly the same categories, or may be interpreting the categories differently.
- − Statistics can be issued selectively to make a particular (usually political) point.
- − They tell us 'what', but not usually 'how' or 'why', something occurs.
- − Statistics are social constructs that reflect the ideas and ideologies of those constructing them.

Documents

These can be divided into public and personal documents, although there is some overlap between the two. Documents are often used by sociologists investigating the past. If it is the relatively recent past, documents may be supplemented by the use of oral accounts or life histories from those who lived through the events. Public documents refer to such things as official reports of the day, treaties, parish registers, court records and letters of state, such as between diplomats and the court, and also novels. Personal documents refer to private letters, diaries, wills, school reports, autobiographies (though if these are published they are also 'public documents') and so on.

Key concepts

Public documents; personal documents; authenticity; representativeness; subjectivity.

Evaluation

- + Historical documents are often the only way of finding out about the past.
- + Personal documents can give a vivid picture of 'what it was like to be there'.

+ Some documents allow researchers to 'see' the same event from two or more different perspectives.
− Personal documents are inevitably subjective, and may even have been written to put across a particular view (e.g. diaries kept by politicians).
− In most cases, there is no way of knowing how representative personal documents are — for example, in the past many people were illiterate and unable to keep diaries or write letters.
− Those historical documents that survive may not be representative.
− The interpretation of documents is inevitably subjective.

The relationship between positivism, interpretivism and sociological methods

Positivism

Positivism is a view that originates from the belief of the natural scientists, particularly of the nineteenth century, that the task of scientists was to discover the 'laws' governing the natural world. Early sociologists, such as Durkheim, believed that human behaviour was similarly governed by 'laws' that the sociologist could discover. Thus, whatever phenomenon is being studied, it should be studied 'scientifically'.

In scientific research, this meant that research should test a hypothesis and should be carried out under controlled conditions, such as in a laboratory. It should be conducted in a systematic manner, with the meticulous and accurate recording of observations and/or measurements, and the conclusions should arise logically from the data generated during the research. All of these factors are intended to reduce 'bias': that is, deliberate or accidental deviation from 'the truth'.

Positivist approaches in sociology favour methods that result in quantitative data, and that can be replicated by other researchers to obtain the same, or very similar, results. This 'hands-off', 'scientific' approach, it is claimed, will help to ensure that the results are free from bias, i.e. are 'objective'.

Positivist-type methods, therefore, mainly involve large-scale social surveys using representative samples, the use of official statistics, and structured interviews. Positivists believe that, properly used, such methods can be objective, value-free and reliable.

Interpretivism

Interpretivism, on the other hand, starts from the assumption that the study of human society and human beings is of a different order from the study of the phenomena of the natural world. Hence, research objectives and techniques that are appropriate to the natural sciences are not necessarily the best to use in sociology. Interpretivists are interested in how people make sense of the world around them, and in discovering the motives that lie behind people's actions. The fact that researchers are humans studying other humans is, for interpretivists, a strength rather than a weakness, as it gives them insight into other people's behaviour — something referred to by Weber as 'empathy' or '*verstehen*'.

Interpretivist-type research methods are those in which people can express their own thoughts and opinions, or can be observed going about their daily lives. They include unstructured interviews, some types of observation and participant observation, and they generally produce qualitative data. For interpretivists, such methods are high in validity. Interpretivists also reject the positivist claim that quantitative methods are objective and free from bias, pointing out the hidden biases in questionnaires and the subjective nature of the collection and presentation of statistics.

Considerations influencing research

A number of factors influence research at all stages. These factors can be divided into theoretical, practical and ethical.

- **Theoretical factors** primarily concern whether the researcher takes a broadly positivist or interpretivist approach. However, it should be noted that much research uses a combination of positivist and interpretivist methods, so the distinction is sometimes a rather artificial one.
- **Practical factors** cover a wide range, including what is to be studied, how much time the researcher has to conduct the research, how expensive it will be, where the funding will come from, how difficult it is to gain access to a group or to particular sources of secondary data, whether the researcher is working alone or can recruit a team of people (obviously this is linked to funding) and the purpose to which the research is to be put. Is it, for example, a postgraduate study to gain a doctorate, or research to inform an aspect of social policy? The researcher's own characteristics also play a part — their gender, age, class and ethnic background could all facilitate or rule out certain types of research, or research into particular groups.

- **Ethical factors** concern whether or not the research deceives the subjects of the study, whether it will cause them any physical or emotional harm, whether it is unjustifiably intrusive into their private life, whether their anonymity can be guaranteed, and also whether the researcher is in danger of physical or psychological harm.

Another influence is the researcher's own interests and concerns. Peter Townsend, for example, feels strongly about the injustice of the unequal distribution of wealth and income, and the social and individual consequences of poverty. He has therefore built his academic life around research into these areas.

Questions
&
Answers

This section of the guide provides you with three questions on the topic of Sociological Methods in the style of the AQA examination, together with two student responses to each question. However, it is important to note that even the grade-A answers are not 'model answers' in the sense that they do not represent the only, or even necessarily the best, way of answering those particular questions. It would be quite possible, particularly in the answers to the two long (e) and (f) questions, to take a different approach, or to use different material, or to come to a different conclusion, and still gain very high marks. The answers represent a particular 'style' of answering the questions: that is, one which answers the question set and displays the appropriate skills, including using appropriate concepts and studies, displaying a critical and evaluative awareness towards the material used, and presenting a logically structured argument.

Examiner's comments

The candidate answers are accompanied by examiner's comments. These are preceded by the icon *e* and indicate where credit is due. For the grade-A answers, the examiner shows you what it is that enables the candidates to score so highly. Particular attention is given to the candidates' use of the examinable skills: knowledge and understanding, and analysis and evaluation. For the grade-C answers, the examiner points out areas for improvement, specific problems and common errors. You might consider rewriting the answer in order to gain higher marks.

Sociological methods I

Item A

For sociologists wishing to conduct ethnographic studies, an important choice is that of which group to study. Unfortunately, this decision does not always lie in the hands of the researcher. For example, some groups have the power to be able to refuse sociologists access to their members. Even if access is granted, such groups are still able to control exactly who and what the sociologist is able to study. It is therefore not surprising that there are very few ethnographic studies of powerful groups in business, the military or politics.

Whatever topic a sociologist chooses to study, a decision has to be made regarding the most suitable method. Again, this choice is influenced by a number of factors.

Item B

A social survey is a method of obtaining a great deal of information from a large number of people, usually in a relatively short time. Having decided on the topic, the sociologist must decide on the population to be studied, and the size and nature of the sample. A key step is the design of the standardised questionnaire and the type of question to be used. It is customary then to conduct a pilot study to test it. As large-scale surveys generate a large quantity of data, analysis is often carried out by electronic means such as optical mark readers, and the results are usually presented in the form of numerical data such as graphs and tables.

(a) Explain what is meant by 'ethnographic studies' (Item A, first line). (2 marks)

(b) Suggest two groups of people, other than those mentioned in Item A, to whom a sociologist might have difficulty in obtaining access for the purpose of conducting research. (4 marks)

(c) Suggest three problems that sociologists might face when carrying out covert participant observation. (6 marks)

(d) Identify and briefly explain two factors that may influence a sociologist's choice of method. (8 marks)

(e) Examine the reasons why some sociologists choose to base their research on secondary sources. (20 marks)

(f) Using information from Item B and elsewhere, assess the usefulness of large-scale social surveys in sociological research. (20 marks)

Total: 60 marks

Answer to question 1: grade-A candidate

(a) Ethnographic studies are in-depth studies of the way of life of a particular group of people, usually using observation or participant observation.

> 🖉 The candidate gives a concise and accurate explanation. **2/2 marks**

(b) People in prison

Judges

> 🖉 This shows how, in this type of question, there is no need to write down more than what is asked for — in this case, two types of group. **4/4 marks**

(c) One problem is that it is difficult to write up any notes, as the research is secret, and the sociologist could easily forget things by the time he or she was able to write things down.

Another problem is that the sociologist can't ask too many questions, or the people would get suspicious, so he or she might not really understand everything that was going on, and could misinterpret things.

A third problem is that the method is considered unethical, as the people being studied haven't given their informed consent to the research. This means that the sociologist would have to think hard about whether the research could be justified, and other sociologists might criticise it.

> 🖉 Again, the candidate has clearly identified three problems, and also made it clear that the problems relate to covert participant observation. **6/6 marks**

(d) One factor would be how much time the sociologist had to carry out the research. Some methods, such as in-depth participant observation, usually take a long time, often several months or even years. If the sociologist didn't have that much time, he or she would need to think of a different method or even a different topic.

A second factor would be whether the sociologist wanted mainly quantitative or mainly qualitative data. If the research was mainly factual, then a method giving quantitative data would probably be best, such as a large-scale survey. However, if the sociologist was interested in people's feelings, or reasons for doing certain things, then this would probably be qualitative data, so a method such as semi-structured or unstructured interviews would be better.

> 🖉 The candidate has identified two factors (time and the type of data required), therefore gaining 2 + 2 marks, and has given a brief and accurate explanation of each, gaining a further 2 + 2 marks. **8/8 marks**

(e) Secondary sources are sources of information that already exist, and that the sociologist hasn't got to find out for himself or herself. These sources include things like official statistics, newspaper reports, personal diaries and letters and minutes of meetings.

📝 The candidate makes a good start to this answer, with an accurate definition of secondary sources and some relevant examples of these.

One reason that a sociologist would use secondary sources is if the research was about things that happened a long time ago, and there was nobody left alive to be able to talk about it. An example of this was when Ann Oakley looked at the role of 'housewife' in historical times, to find out the kinds of things that women did centuries ago. Even if there were still people alive to ask, the sociologist still might want to use historical sources to compare with the people's own reports, for example to see whether written accounts of poverty in the 1930s were the same as those of people who had actually experienced this poverty.

📝 A valid reason for using secondary sources is identified and further explained.

Another reason for using secondary sources is that the method requires it. In content analysis, for example, the method uses secondary data such as newspaper reports or soap operas to see how much space or time is given to certain things, or how a particular issue is treated, or how it changes over time. You can only use this method with secondary sources.

📝 This is a very good reason, and not perhaps one that immediately springs to mind. As in the previous paragraph, the candidate not only identifies a reason, but writes about it in a way that shows a clear understanding.

Another reason is that the sociologist might not be able to get the information any other way. If you wanted to know what proportion of the people in different parts of the country had incomes below the national average, or were unemployed, or how many crimes were committed per head of population in different areas, then it would be very difficult and time consuming for the sociologist to try to find this out. However, the figures would be available in official statistics. Again, if a sociologist wanted to know how many pupils were being excluded from school, or what proportion of schools in an area were at the top or at the bottom of the league tables, then the information would be available in secondary sources.

📝 A further accurate reason is given, supported by clear and accurate examples.

Another reason would be to make comparisons over time, for example by comparing certain information from the latest census with the last one, to see what changes had taken place over ten years.

📝 This is similar to the reason given in the first paragraph, but uses a different example and draws out the idea of using secondary data to make comparisons.

Finally, a sociologist might use secondary sources because of the belief that secondary data are more objective and truthful and contain less bias than material collected first-hand by the sociologist. Positivist sociologists would be likely to take this view, whereas interpretivist sociologists would disagree as they point out all the problems with secondary data. They would be more likely to choose primary data.

question

📝 This reason draws more on theoretical than practical reasons, and makes a comparison between positivist and interpretivist sociologists. It would have been helpful to have identified some of the problems that interpretivist sociologists believe to exist with secondary data, rather than just stating the point.

This is a well-constructed and relevant answer that is clearly focused on the question. The points made are accurate and are well supported by examples, although most of these are general, rather than identifying specific pieces of research. The knowledge and understanding are good, and the answer has an analytical structure. However, there is little evidence of the skill of evaluation. Even though 14 of the available marks are for knowledge and understanding, 6 marks are for the skills of identification, analysis, interpretation and evaluation. It is important to try to make some evaluative points in (e) answers. **15/20 marks**

(f) As Item B says, social surveys allow the sociologist to obtain information from a large number of people fairly quickly. This makes it a useful method for the kind of research that needs a lot of straightforward information rather than an in-depth look at something.

📝 This immediately uses material from Item B, but goes further than simply repeating phrases from the item. The second sentence makes an evaluative point about the conditions under which social surveys can be useful.

There are different types of social survey. Some use postal questionnaires, while others use interviewers to ask the questions, or at least be present while the respondent answers them. Some use almost all closed questions (questions with yes/no or a pre-determined set of answers) while others use a mixture of closed and open-ended questions (questions where the respondent is able to answer in his/her own words). Some surveys use straightforward random sampling, while others use more sophisticated methods, such as stratified random sampling or multi-phase sampling. The type chosen will depend on the kind of information needed and the characteristics of the population involved. Some sociologists, e.g. positivists, would choose large-scale surveys because they think this method is more reliable and not as open to bias as more interpretative methods such as unstructured interviews and participant observation studies.

📝 This paragraph shows good and accurate knowledge and understanding of different types of questionnaire and sampling methods, but probably gives more information than is necessary. The last sentence, however, is important, and could have been developed to explain further the different views on surveys held by positivist and interpretivist sociologists — in particular, why positivists might think this method is 'more reliable and not as open to bias' as some other methods.

Again, sociologists can make use of social surveys as primary or secondary data. Some secondary data social surveys of use to sociologists would include the national census, the British Crime Survey and the British Social Attitudes Survey. None of these could really be carried out by an individual sociologist, as they would be too expensive and time consuming. Sociologists who have used their own social surveys include Townsend on poverty.

🖉 The candidate makes a good point — namely, that social surveys can use either primary or secondary data — and gives appropriate examples.

Social surveys can be of great use to sociologists. The large-scale government surveys give them information they would not be able to get on their own, and which could allow them to develop a hypothesis to test. Large-scale surveys allow comparisons to be made between different parts of the country, for example you could compare opinions on asylum seekers in areas where there were a lot of them, with areas with none. Again, large-scale social surveys allow a big population to be studied relatively cheaply and, if necessary, in a short space of time. This can be important with issues like the general election, when it is important to know how people's opinions are changing as the campaign takes place.

🖉 This is an evaluative paragraph explaining how and why surveys can be of use (to make comparisons and to save time), with a relevant example of each point.

However, there are drawbacks with this method. Even with standardised question-naires, interviewer bias can creep in, as respondents can relate, even uncon-sciously, to the social characteristics or tone of voice of the interviewer. Closed questions do not allow respondents to give reasons for their answers, and the answers provided may not quite fit what they want to say. Postal questionnaires have a low response rate, and with no-one there to answer questions, respon-dents may be confused and not understand the questions properly. They can even get someone else to fill it in for them. If the survey sample isn't chosen carefully, the findings will not be representative.

🖉 This evaluative paragraph makes a number of points about the potential disadvan-tages of large-scale social surveys.

However, sociologists are influenced by many different factors when choosing a method, and in some cases the large-scale social survey will be thought to have more advantages than disadvantages for a particular piece of research. It really depends what it is that the sociologist wants to find out.

🖉 This concluding paragraph draws attention to the fact that the choice of method depends on a number of factors, and that any method should really be judged against the particular research for which it is used.

There is a logical structure to this answer, and knowledge and understanding are relevant and sound. Examples are well applied to the question. More use could perhaps have been made of the item: for example, developing the points about questionnaire design and the analysis of data. It would also have been helpful to say something about reliability and validity in the discussion of different views on the usefulness of large-scale surveys. There is evidence of evaluation in the identification of the advantages and disadvantages of the method, though this is somewhat 'list-like'. However, the answer meets the criteria of the top band, although at the bottom rather than higher end of the band.   **15/20 marks**

Overall: 50/60 marks

■ ■ ■

Answer to question 1: grade-C candidate

(a) Ethnographic studies are when a sociologist studies a gang or something, like James Patrick in Glasgow.

> 🖉 While the candidate gives an *example* of an ethnographic study, the term is not explained.   **1/2 marks**

(b) A sociologist would have difficulty in getting access to doctors, because their records are confidential and they wouldn't want to talk about their patients, and also to people like Tony Blair, as he would be too busy.

> 🖉 Note that the candidate has written more than is necessary — simply to name two suitable groups would be sufficient. While the first group, doctors, gains 2 marks (although the candidate was not asked to give reasons), the second example does not score marks, as powerful groups 'in politics' was one of those mentioned in Item A. Always read the question carefully to see whether certain examples would be excluded.   **2/4 marks**

(c) Covert participant observation is a good method used by interpretivist sociologists. Positivist sociologists criticise the method because they say it is biased. Also, there is the difficulty that the sociologist could put themselves in danger if the group was engaged in illegal activities like robbery or drug-taking, and people can act differently if they know they are being watched.

> 🖉 This is not a good way of laying out the answer to a question such as this. Everything is contained in a single paragraph, rather than being shown as three separate points. However, examiners will always read carefully what a candidate has written, to make sure that correct answers are rewarded. Here, the candidate did not need to write the first sentence — this was not asked for. What problems associated with covert participant observation has the candidate identified? Firstly, we can see that the candidate is trying to make the point that the research findings might be criticised for being biased, although we are not told why. Even allowing for 'benefit of the doubt', the candidate has not really given an example of a problem faced by sociologists when carrying out this method. The second point scores the marks, for identifying the risk of some kinds of covert participant observation. Sadly, the third point does not score, as the candidate has lost the focus on *covert* participant observation — what has been identified is a potential problem with observation or overt participant observation.   **2/6 marks**

(d) One factor is whether the sociologist is a positivist or an interpretivist. Positivist sociologists would use surveys, as they think these are scientific, while interpretivists would use unstructured interviews or participant observation because they want to study people's meanings.

Another factor is time, as some methods take a long time to carry out.

✍ The first part gains 2 + 2 marks, as a factor is identified (taking a positivist or interpretivist approach) and this is explained, although not very fully. The second part gains 2 marks for identifying 'time' as a factor, but it does not really give an explanation. The candidate should have explained that in some cases it would be possible to spend a long time on a piece of research, but there are other cases in which the research needs to be completed within a particular time span — either because there is a limited budget or because the findings are needed fairly quickly. So, 2 + 0 marks are gained for this part of the answer. **6/8 marks**

(e) Secondary sources can be useful to sociologists, especially official statistics. These are available on things like crime, divorce, GCSE passes and so on.

✍ Here the candidate is 'jumping straight in' rather than giving an introduction to the answer. Nevertheless, the point is made that secondary sources can be 'useful', and an example of a particular type of secondary source is given.

Using official statistics can save the sociologist a lot of time and money. For instance, if the sociologist wants to do some research on divorce, it would be hard without the divorce statistics to know how many people are getting divorced and whether this is going up or down. This is also true of exam passes, as the sociologist would otherwise have to go to every school and college and ask them for their pass rates, and many of them wouldn't want to give out this information.

✍ A reason for using official statistics is provided, together with some discussion. Note that so far the candidate has confined the discussion of 'secondary sources' to official statistics.

Another reason is if the sociologist is a positivist, then he would use official statistics because these give numbers and figures, and positivists believe that this makes the information more reliable.

✍ Another appropriate reason is provided, although the supporting information is rather simplistic. Although the point about reliability is a good one, the candidate doesn't make it clear that 'reliable' is being used here in the sociological sense of the word. Note also the mildly sexist language in that the sociologist is referred to as 'he'. It is best to avoid this if possible.

Sometimes a sociologist would use official statistics to get some information but then go and use another method as well. For example, he could use the statistics to see that ethnic minority children didn't do as well in exams as white children, and then go and do some interviews with teachers or pupils to try to find the reasons for this.

✍ This is a good and useful point — namely, that sometimes the use of secondary data is complemented by other methods. A relevant example is given, but note that the phrase 'ethnic minority children' is not a good one to use in this context, as the statistics show that it is children from some ethnic minority groups who fail to do as well as they might. Note that the focus is still only on official statistics — the candidate has so far failed to show knowledge of any other kind of secondary data.

Sociologists who want to study things in the past will have to use secondary data because the people aren't alive to interview. They can use things like letters and diaries or parish records, like Laslett and Anderson in their study of family structures in the past. Also, we have proof that the birth rate and the infant mortality rate have fallen because we have the statistics to show what these were in the past. The figures show us 'what' but not 'why'.

✍ The candidate gives another reason for using secondary data and (at last!) mentions some other types of secondary source. Note that this candidate makes the common mistake of implying that 'Laslett and Anderson' carried out a single piece of research together, whereas they worked independently and used different sources of data. However, no penalty is imposed for this, as marking is always positive. A further potentially important example is given, i.e. birth and infant mortality rates, but again, this is not developed. The last sentence in the paragraph could certainly have been used as the basis for making important points about some of the problems of interpreting statistics.

However, there are lots of problems with official statistics, and sociologists have to be careful if they are going to use them, as they can give a false impression.

✍ There is an element of evaluation here, i.e. the recognition that official statistics have 'lots of problems', although the candidate fails to capitalise on this point by saying what these are, or why they might give a 'false impression'. Note that we are back again to official statistics, and the candidate has lost the opportunity to comment on some of the problems of using historical data.

This answer matches the mark band descriptor of 'a reasonable knowledge and understanding of some potentially relevant material', which puts it in the 8–14 mark band. The candidate loses the opportunity to widen the discussion because of an over-reliance on official statistics as sources of secondary data. More studies could have been quoted to support some of the points being made. However, the material presented, although limited, is generally accurate, and one or two evaluative points are made, although these are neither explained nor developed.

10/20 marks

(f) As Item B says, a social survey is a method of obtaining a great deal of information from a large number of people, usually in a relatively short space of time.

✍ While it is important to 'use the item' when instructed to do so, simply copying out chunks will not gain marks and is a waste of time. The information has to be 'used' in some way, not just copied!

One of the largest social surveys is the Census, which asks questions about every household in the country and is carried out every 10 years. This provides sociologists with a lot of very useful information, and they can look back at earlier Censuses to see what changes have happened in the meantime.

☑ An appropriate example is identified and explained, and its usefulness to sociologists is mentioned, though not really developed.

There are lots of steps to go through when designing a social survey. As it says in the item, you must choose the topic, and then the type of sample. Samples are used so you don't have to survey everybody in the population, you can just ask the people in your sample, and if it is representative, you will get the same results. You also have to design your questionnaire and decide whether you want open-ended or closed questions. You should then do a pilot study to check that everything is all right, and people understand and can answer the questions. You also have to decide whether you are going to use interviewers or send your questionnaires through the post (postal questionnaires).

☑ The candidate shows a reasonable knowledge of social surveys, and this time develops some of the points from the item, rather than just copying them.

Sociologists have used social surveys for things like finding out whether people have committed crimes or have been the victims of crime, whether they have truanted from school, and what things they think people should be able to afford so as not to be in poverty. Social surveys are also used a lot when there is a general election, to see how people are going to vote, or to find out what people think about an important issue like giving fathers paternity leave or lowering the age of consent for male homosexuality.

☑ The candidate gives a comprehensive list of situations in which social surveys have been used. At this point it is simply a list, and no comments have been offered on whether there are differences in the different types of survey mentioned. What kinds of points do you think that you could make about the examples given above?

Social surveys are useful because they can give the sociologists a lot of information. If the questions are mainly closed, then the information can be analysed quite quickly and the results known in a short space of time. This is important if the survey is on how you are going to vote, because if it takes too long then the election will be over. If you have a team of interviewers, or if the questionnaires are sent though the post, then it is easy to get answers from people in different parts of the country. This could be important if you wanted to compare things, say how long people were waiting for hip operations in different places. It is also a good thing because you can show your results in graphs and tables that people can understand quickly, without having to wade through a whole book to find out.

☑ A number of relevant points are made here, still mainly focused on the advantages of large-scale social surveys. Remember that to 'assess' the usefulness, some disadvantages or possible problems will also have to be discussed.

However, social surveys have problems too. Postal questionnaires have a low response rate — that is, not many people return them. With interviewers, you can get interviewer bias, which affects the results. Some people won't understand the questions, and will give the wrong answer, and some people won't be able to find an answer which suits them, so will have to choose something else.

e A rather hasty mention is made of some of the problems of surveys, with a tendency just to state the problem rather than explain it.

Still, as I have said, social surveys can be useful and sociologists will continue to use them.

e This kind of conclusion doesn't add anything to the discussion and is best avoided.

This candidate shows a reasonable knowledge and understanding of social surveys, and there is some limited evaluation, which is not really developed. The material offered is interpreted in a broadly sociological manner. To gain higher marks, there could have been a discussion of the concepts of reliability and validity, and how these relate to large-scale social surveys, and more about the factors that influence the use of large-scale social surveys. **11/20 marks**

Overall: 32/60 marks

Sociological methods II

Item A

Many sociologists choose to use secondary sources of data when conducting research. For some areas of research there is a wealth of published information which sociologists can use. For example, government agencies produce official statistics covering a wide variety of topics, and many non-government organisations also produce data of interest to sociologists. Some sociologists, particularly those conducting historical research, make use of a variety of personal documents.

In an article on using documents in sociological research, John Scott suggests that to judge their usefulness, the sociologist should look critically at their authenticity, credibility, representativeness and meaning.

Item B

Unstructured interviews are usually chosen to allow the sociologist to explore particular issues in depth. While they may be group interviews, with several respondents taking part at once, it is more usual for such interviews to be carried out with just one respondent. If the research is about marriage or aspects of family life, sometimes both spouses or partners will be interviewed together.

Conducting a successful unstructured interview is a very skilled process, as the interviewer must put the respondent at ease and gently guide the discussion towards the topic or topics of interest, while at the same time not 'prompting', and allowing the respondent to raise other aspects which the researcher may not even have considered. Unstructured interviews are usually recorded, to make the situation more like a normal conversation and also to allow the interviewer freedom to observe the facial expressions, body language and tone of voice of the interviewee, as these can be as revealing as what is actually said.

(a) **Explain what is meant by 'representativeness' (Item A).** (2 marks)

(b) **Give two examples of 'personal documents' that might be used by sociologists (Item A).** (4 marks)

(c) **Suggest three reasons why personal documents may not be representative.** (6 marks)

(d) **Identify and briefly explain two problems faced by sociologists when using official statistics.** (8 marks)

(e) Discuss some of the ways used by sociologists to draw a sample. (20 marks)

(f) Using information from Item B and elsewhere, assess the usefulness of unstructured interviews in sociological research. (20 marks)

Total: 60 marks

■ ■ ■

Answer to question 2: grade-A candidate

(a) Representativeness means that something is the same as or very like other things of the same type, and not something unusual or one-off. A representative sample is one that gives the same or similar results to what you would get if you surveyed the whole population. A representative document would therefore be a typical example of other documents of that kind.

✍ The candidate has taken pains to ensure that the examiner recognises their under-standing of the concept by giving more information than is necessary. While this is not a good technique if you are confident that you have answered the question in just one sentence, if you are not quite sure that you have made your under-standing clear, it is sometimes helpful to give a little extra information, perhaps in the form of an example. (Remember, though, that an accurate example without any definition will score only 1 of the 2 marks.) **2/2 marks**

(b) School reports and diaries.

✍ Two appropriate examples are given. Can you think of any other examples that might apply only to the electronic age? **4/4 marks**

(c) One reason is that written personal documents like diaries and letters can only be produced by people who can read and write, so poor people aren't likely to be represented, as until quite recently they were illiterate.

Another reason is that many documents are lost or destroyed and only some survive over time, and there is no way of knowing whether they are typical of other similar types of document from that time.

A third reason is that there is no way of checking whether the writer was telling the truth when s/he wrote in a diary or letter, and the sociologist only gets one side of the story.

✍ Three clear and accurate reasons are given, showing a clear understanding of some of the problems of 'representativeness'. This candidate has chosen to focus on historical documents — do you think that the reasons given would all apply to contemporary personal documents, including any you might have identified as a possible answer to question (b)? **6/6 marks**

(d) One problem, especially when using 'soft' statistics, is that statistics are a social construct. This means that what counts as a particular statistic depends on the interpretation of the person compiling them, and the sociologist has no way of knowing what the person chose to count 'in' or leave out, as this could have been a personal decision.

Another problem is that it is difficult to compare statistics over time. For example, it is not easy to compare the school performance of pupils over the past 50 years, as there have been changes in the education system, especially the move from the tripartite to the comprehensive system and various changes in the exams pupils could take, as well as the raising of the school leaving age. Pupils who went to secondary modern schools usually left at 15 and didn't take any exams, so the percentage of pupils with O-levels at that time was very small, but this didn't mean that children were less intelligent.

> Two reasons are clearly stated and each of them explained, so 2 + 2 + 2 + 2 marks are gained. Note that in the first example the examiner is able to see quite clearly that the candidate understands what is meant by a 'social construct'. This is then applied to the question. The second example shows how good candidates are able to bring in material from another part of the course when answering questions about sociological methods. **8/8 marks**

(e) Sociologists use a sample when they want to do research on just a selection of people from a wider population. In most cases, sociologists want to make their sample representative, so that the findings from their research on the sample will be able to be applied to the whole population. In some types of research, this is not possible.

> This is a good introduction. Note that the candidate doesn't just 'dive in' to the question, but takes the time to explain why sociologists would want to draw a sample. This is likely to be of benefit to the rest of the candidate's answer.

A very common sampling method is random sampling, sometimes known as probability sampling, where everybody in the population has an equal chance of being selected. Each unit of the population (e.g. each person in a town) is given a number, and the amount of people needed for the sample are picked out using special random number tables. However, sociologists will sometimes use a 'sampling frame', which is a list of people, like the electoral roll of all the people in a town. They can then pick out, for example, the 4th, 14th, 24th, 34th, etc. person on the list until they have got the number of people they want. School registers can be used in this way for research in schools.

> The candidate gives a clear description of two ways of drawing a random sample, using appropriate concepts, and also an example of a situation in which a particular type of sampling frame is often used.

Another type of sampling is quota sampling. This is where the sociologist has a clear idea of what characteristics they want people in the sample to have, and

what proportion should have them. For example, the sample might need equal numbers of males and females, a certain number of people in different age groups, or different social classes, or different ethnic groups, etc. You obviously need to have all this information about the people in your sample, or else you have your interviewers standing on the street picking out people who seem to have the characteristics needed until the quota is filled. This can be quite difficult, especially with things such as social class.

Again, the candidate gives a clear and detailed description, showing a good understanding of another type of sampling procedure, and gives an evaluative point at the end.

There is snowball, or opportunity, sampling. This is used where the sociologist hasn't got a sampling frame or any kind of list of the people wanted for the research. This can be because they are deviant, such as criminals or drug-takers, or because there is no need for anybody to have such a list, like the divorced people under 40 in a particular town, or because there is a list but it is confidential, such as people who are HIV positive, and the sociologist wouldn't be able to get hold of this. What happens is that the sociologist needs at least one person who fits the bill. This person is then interviewed and asked for the name of at least one other person like themselves. This person is interviewed and asked for another name or names, and so it goes on until the sociologist has enough people for the research. This kind of sampling, though useful for some kinds of research, is not as reliable as random sampling, because the way the people are chosen is biased — they may not be like others in the group, as it all depends on the names people are willing to give the sociologist.

This is another clear discussion of a sampling technique which brings in an important evaluative point at the end about reliability. Remember that although (e) questions are weighted towards knowledge and understanding, 6 of the available marks are for identification, analysis, interpretation and evaluation, so you must be sure to make at least a few evaluative points.

Finally, there is a self-selected sample. This is when the sociologist asks or advertises for people who have had certain kinds of experiences, or who fit a certain category, to volunteer to take part in the research. This would be used when it would be difficult or impossible for the sociologist to find out who these people were, such as people who had been evacuated as children in the war, or who lived on a particular housing estate at a particular time. While obviously useful, and perhaps the only way for some kinds of research, this is another method where the people who come forward are not necessarily representative of all the people in that group. This kind of sampling is usually used when the sociologist is going to do in-depth research.

Again, this is a clear and accurate discussion of another sampling technique, showing clear understanding and the ability both to use relevant examples and to make an evaluative point.

The sampling method chosen by a sociologist will depend on the type of research and in particular on the availability or otherwise of a suitable sampling frame.

🖉 This is a brief conclusion that nevertheless reinforces two important aspects of sampling discussed in the main body of the answer.

This answer meets all the criteria of a top-band answer to an (e) question. The knowledge is sound and there is good use of appropriate sociological concepts. The material used is focused on the topic of the question, and the answer is well organised with a good logical structure. A number of evaluative points are applied to the question and to the material presented in answer to it.

20/20 marks

(f) As it says in Item B, unstructured interviews are usually chosen to allow the sociologist to explore particular issues in depth. Sociologists are able to do this because unstructured interviews do not have lists of questions, or interview schedules. Each one is like a conversation between the interviewer and the respondent. Obviously, the interviewer has an idea of the topics s/he wants to cover in the interview, and may introduce these by asking questions such as 'Have you ever...?' or 'What did you think about...?' and so on. However, the respondent is able to answer freely, and may say things that take the interview in a new direction, perhaps introducing something the sociologist hadn't even thought of, again as it says in the item.

🖉 The candidate makes immediate reference to Item B. While it is not usually a good idea simply to repeat the words used in the item, the candidate goes on to explain and develop one of the phrases quoted, about exploring particular issues in depth. Remember that all the information given in an item is potentially useful in answering the question, so don't stop just because you have mentioned the item once. At the end of the paragraph, the candidate makes reference to another aspect of unstructured interviews from the item.

Unstructured interviews are very useful for certain kinds of research, where the sociologist is interested in finding out about people's meanings and motives or about the effect of something on their life. Examples of research where unstructured interviews would be useful are sensitive areas such as domestic violence, rape or racism, or a topic where the people involved haven't been studied much or at all, like housewives or new mothers. They are also useful for research into deviant groups, where the subjects wouldn't be likely to answer questions on a questionnaire or if they were just stopped on the street.

🖉 This is quite a good paragraph outlining the kinds of situation where unstructured interviews would be useful. However, it would have been even better if the candidate had mentioned some specific pieces of research in which unstructured interviews had been used. Can you think of some, perhaps using the list of situations given in the paragraph above?

Another advantage that unstructured interviews have over questionnaires is that as they are usually recorded, the researcher is free to pick up other clues, such as those mentioned in Item B — expressions, body language and tone of voice. These can be important, because people's words alone don't always give a true impression of what they are feeling. For example, the researcher could notice that the person looked angry when talking about a particular person or thing, even though their words weren't angry, or that they had very defensive body language even though they weren't saying anything about being upset or feeling threatened. As it is being recorded, changes in tone of voice can be picked up later, even if they weren't noticed at the time.

✐ The candidate again makes reference to Item B by quoting from it, but develops and gives examples of the points made, rather than simply letting the quotes speak for themselves.

However, it is important to remember that unstructured interviews also have disadvantages. The interviewer has to be very skilled, as it says in the item. The respondent has to feel at ease and has to trust the interviewer, and forget the tape recorder and come to feel that they are taking part in a conversation, rather than being interviewed about something. Otherwise they may not tell the truth, or may hide certain things which could be very important. So, unstructured interviews will not work well unless the interviewer has these important skills.

✐ This is an important paragraph as the candidate here moves to a more critical view of unstructured interviews. In order to 'assess' something, you have to show more than one side of the argument and be able to identify both strengths and weaknesses.

A more crucial disadvantage is that in most research using unstructured interviews not many can be carried out, as they are very time-consuming and therefore expensive. This means that their reliability can be questioned — other researchers carrying out interviews with the same kinds of respondents may get different results. It is not easy to generalise from unstructured interviews. Also, the way that the sociologist reports the findings may be biased, as s/he will have to select from all the material they have recorded and pick out the things they think are important. This is very subjective, and another sociologist listening to the tapes might focus on different things and come to a different conclusion. Again, unless someone else listens to all the tapes, we do not know whether the sociologist was guilty of 'prompting' the respondent to talk about something they may not have mentioned. As Item B says, you shouldn't 'prompt' in this kind of research, as this is another way of introducing bias into the research.

✐ Here are more potential criticisms of unstructured interviews, introducing some very important concepts — reliability, bias and subjectivity. It is always important to include relevant sociological concepts in your answers, as this will show that

they are written from a sociological viewpoint and are not just 'common sense' or things that could have been written by someone who had not studied sociology.

However, sociologists who use unstructured interviews defend them by saying that they are a valid method — that is, they feel that such interviews are the best way of getting at the information they want for their research, and give better quality information than, say, questionnaires. In some cases, sociologists try to overcome the possible drawbacks of unstructured interviews by combining them with another method of data collection, such as using official statistics or question-naires, so that they can compare the information. There is no doubt that unstruc-tured interviews are a useful and important method, and sociologists are likely to go on using them in cases where they think that they represent the best way of getting the information they want. They just need to keep aware of the possible problems and try not to let them cause bias.

In this paragraph the candidate tries to sum up the argument by introducing the counter-claims to the criticisms of unstructured interviews and making reference to another important concept in this debate — validity. The important point is also made that some sociologists combine the use of unstructured interviews with other methods, and a reason for this is given.

This is a competent account of some of the main strengths and weaknesses of unstructured interviews. It is clearly written and well laid out, and shows good sociological knowledge and understanding of concepts important to this debate. What is lacking is any reference to the theoretical perspectives involved. The discussion of reliability and validity would have benefited from appropriate mention of how these are linked to sociologists taking a broadly positivist or interpretivist approach. The conclusion is a little weak — an assertion is made that unstructured interviews represent 'a useful and important method', but it would have been a good idea to remind us exactly why this is the case. Another weakness in the answer is that no reference is made to any sociological research that has used unstructured interviews. In any answer on a particular method, always make sure that you are able to mention at least one study that has used the method, and preferably more than one. Nevertheless, this is a good answer that shows evidence of the AO2 skills and keeps a focus on the question. Very good use is made of the item. **15/20 marks**

Overall: 55/60 marks

■ ■ ■

Answer to question 2: grade-C candidate

(a) Representativeness is when something is representative of something else.

This does not explain at all what is meant by 'representativeness', so no marks can be awarded here. This shows the importance of learning the meanings of all the sociological concepts that you meet during the course. **0/2 marks**

question

(b) Personal documents are things that are very personal to you, like photographs or love letters.

 The first phrase is unnecessary, but two appropriate examples are mentioned, so 2 marks are awarded for each. **4/4 marks**

(c) Personal documents may not be representative just because they are personal. What you write about in your diary, for example, will be what you think and what happens to you, so it may not be the same as what other people think or what happens to them — your life and how you feel will be different. Again, if we take love letters as an example, these would not be meant to be read by anybody else except the person they were sent to, so most love letters would not be kept to be found by sociologists, unless they were sent to somebody very famous like royalty, whose letters were kept. Famous people aren't like everybody else, so their love letters might not be typical.

 This shows the importance of separating out the points you make so that you can be sure that you have given the number asked for. In this example, the candidate makes two relevant points (although in a rather long-winded manner): first, that people's experiences and their reflections on them will differ, so diaries will usually only tell us something about an individual; and secondly, that sociologists will only usually be able to obtain a small selection of certain types of personal document, and again these may, as in the example above, relate to people who are not at all 'typical'. There is no attempt to provide a third reason, and the candidate has written a lot for just 4 marks. **4/6 marks**

(d) One problem with official statistics is that they will have been collected by somebody else, sometimes several other people, and the sociologist won't be able to ask them to explain something. Domestic violence wasn't really regarded as a crime until quite recently, so if the sociologist was looking at crime statistics, he might not know whether domestic violence had been included as a crime or not. Another problem is that sometimes the laws have changed. People used not to be able to get a divorce without a special Act of Parliament, so there were hardly any divorces; then divorce became legal and more people got divorced. When women could get a divorce on the same grounds as men the divorce rate went up again, so unless you knew all this, you could think from looking at the statistics that people used to have much happier marriages than they do nowadays, but you would be wrong.

 Two problems with statistics have been identified and each has been explained using a suitable example, so this answer would receive 2 + 2 + 2 + 2 marks. **8/8 marks**

(e) There are lots of different ways sociologists can draw a sample for their research. These include random sampling, quota sampling and snowball sampling.

 Three appropriate types of sample are mentioned.

In a random sample, all the names of the people in the survey can be put into a hat and then the number that the sociologist needs is drawn out of the hat. Of course, sociologists don't usually use the hat method; instead they give everybody in the survey a number and use special tables to choose the ones in the sample. In a random sample, everybody has the same chance of being picked.

> This is a fairly simplistic description of random sampling, but it makes the important point about the units having the same probability of being chosen.

In a quota sample, the sociologist decides what kinds of people he needs in the sample, e.g. how many men, how many women, what ages they should be, what ethnic group, etc. The interviewers are then each given a quota and told to stop passers by until they have got the right number in their quota. This is not a very good method and is usually used by market researchers who want to know what soap powder you use, rather than by sociologists.

> Again, a basic description of quota sampling is given. It is a pity that the candidate didn't say why this is 'not a very good method', and why it isn't often used by sociologists — this would have shown the skill of evaluation.

In a snowball sample, the sociologist has to find somebody to start the snowball rolling. He has to have one person who will talk to him about the things he wants to know. Once he has found this person, like a member of a gang, or somebody who takes ecstasy, then he can ask them to give him the name of somebody else in the group, and he can go to them and ask them for another name and so on until he has been able to talk to enough people. The hardest part of this is finding the first person and getting them to trust you enough to give you the name of one of their friends.

> This is a reasonable description of snowball sampling, with a couple of examples to show the kind of context in which this method might be used.

Sometimes the sampling doesn't work. In postal questionnaires, for example, the response rate is low, and lots of people don't send back their questionnaires, so the sample isn't as good as it should be, which can affect the results of the research.

> There is potentially an important evaluative point here, which is not developed. The candidate could have made the point much more clearly by using appropriate sociological concepts such as 'representativeness' and 'reliability'. Always try to use appropriate concepts when answering questions. This answer comes to an abrupt close, with no attempt to write a conclusion.
>
> The candidate shows reasonable knowledge and understanding of three types of sample, but does not say why sociologists would wish to draw a sample in the first place. A few appropriate examples are given, though no study is mentioned by name. Points tend to be stated with no wider discussion. There is a broadly logical structure to the answer, but analysis and evaluation are both very limited. **10/20 marks**

(f) Unstructured interviews can be very useful. An unstructured interview is like a conversation, and there is no questionnaire or list of questions. Sociologists use unstructured interviews to find out about things that people would find it hard to talk about in a questionnaire or if talking to a stranger on the street. These include domestic violence and why people belong to a cult.

✏ Some brief points are made about unstructured interviews and the contexts in which they might be used.

As it says in Item B, you have to be skilled to do an unstructured interview, and they also take a long time, sometimes several hours. This means that the sociologist can't do as many interviews as they would if they were giving out questionnaires. This means that the findings are only about a small number of people, not several hundreds like you might get with a questionnaire.

✏ There are some important points here that should be developed. The issue of 'skill' could be explored further, making use of the information from the item. The point about the small number is an important one and provides the candidate with the opportunity to talk about validity as opposed to reliability, but the point is not developed further.

One problem with unstructured interviews is that unless you listen to the tape recording, you have to take the sociologist's word that that is what the people in the interviews actually said — there is no way of checking. The researcher could have just picked out the things she wanted to hear and ignored the rest.

✏ Here is another missed opportunity — this time to talk about the difficulties of checking the data and the possibility of interviewer bias. These potentially important criticisms are left implicit.

Also, some people like positivists are against unstructured interviews as they don't give you tables and graphs, but just people's words and what the sociologist thinks that these show.

✏ This is another potentially important point that is not explained and explored further. Think of how this paragraph could have been developed.

There is evidence of some sociological knowledge and understanding of unstructured interviews, and some potentially important evaluative points are made, but these are not explored or developed. Little use is made of the item, which provides some useful material that the candidate could have developed. Rewrite the answer to this question, making as much use of the item as possible (remember not simply to copy out phrases though) and illustrating your answer with reference to actual sociological studies. In addition, bring in the discussion about validity and reliability, and which type of sociologist would be more likely to make use of unstructured interviews, and why. Remember to include as many criticisms of the method as you can think of, as well as discussing its strengths. **9/20 marks**

Overall: 35/60 marks

Sociological methods III

Item A

While a lot of attention has been given to the difficulties involved in conducting participant observation in sociological research, it is equally true that non-participant observation can also pose problems for the sociologist. This method involves more, of course, than just 'observing'. The sociologist has to make sense of, and interpret, what is seen. There is also the danger of introducing the Hawthorne effect.

One kind of observation that has sometimes been used in classrooms involves the researcher drawing up a list of particular activities or interactions and completing a 'tally chart', in which a tick is made when one of the designated activities occurs during a given period of time. This method produces quantitative data measuring what has been going on during the period of observation.

Item B

The National Child Development Study is a longitudinal study which is following about 17,000 people all born in Great Britain between 3 and 9 March 1958. The aim of the study is to improve understanding of factors affecting people's development over the whole life span. To date, there have been five attempts to trace all the people in the group to monitor their physical, educational and social development. These took place when they were aged 7, 11, 16, 23 and 33. Information was obtained from the mother and medical records from the midwife. This has been supplemented by various other information acquired throughout the period of the study. In addition, when the people in the group were aged 20, contact was made with the schools and colleges which they attended. The information gained from the study is used for a wide range of research, including medical/health research.

(a) **Explain what is meant by the 'Hawthorne effect' (Item A, line 6).** (2 marks)
(b) **State two ways in which quantitative data (Item A, lines 10–11) may be displayed.** (4 marks)
(c) **State three criticisms that could be made of participant observation as a method of sociological research.** (6 marks)
(d) **Identify and briefly explain two reasons why the 'tally chart' method of observing classrooms described in Item A may not give an accurate picture of what is actually happening.** (8 marks)
(e) **Examine some of the problems involved in constructing a questionnaire for use in sociological research.** (20 marks)

(f) **Using information from Item B and elsewhere, assess the advantages and disadvantages of longitudinal studies in sociological research.** (20 marks)

Total: 60 marks

■ ■ ■

Answer to question 3: grade-A candidate

(a) The Hawthorne effect is when people who know they are being studied by a researcher behave differently than they would otherwise. That is, the fact that they are being observed alters their behaviour.

> 🖉 This is an accurate explanation which makes the important point that the people have to know that they are being observed or studied in order for the effect to take place. **2/2 marks**

(b) Quantitative data refers to numerical data, so they could be displayed in a table or drawn as a graph.

> 🖉 The candidate gets full marks for identifying two correct ways in which quantitative data can be displayed, but there was no need to explain the meaning of quantitative data. The answer would have been better laid out if the two ways had been stated on separate lines. **4/4 marks**

(c) If the PO is covert, then it is unethical, as it is wrong to study people when they haven't been able to give the researcher their 'informed consent'.

PO is very subjective; we have to rely on what the researcher tells us happened during the study; it is only their view and someone else could interpret things differently.

As PO is usually used in small-scale studies, it is not easy to make generalisations from it; it really only applies to the group being studied at the time.

> 🖉 The candidate makes three accurate and relevant criticisms of participant observation. Note that as the question asked about 'participant observation', rather than a particular form of it, the candidate carefully made it clear in the first criticism that this applied to covert participant observation. **6/6 marks**

(d) One reason is that the researcher has to have drawn up the tally chart before going in to do the classroom observation. This means that s/he has already had to decide what things to look for. As there is no way of adding things to the chart once the observation has started, some important things could have been left out which will not appear in the research.

Another reason is that this method tells us 'what' but not 'why'. It only describes what happens at certain times, and can't tell us anything about why the pupils and teacher were acting as they did, which might be a very important part of the classroom behaviour. For example, the pupils might only behave that way with a certain teacher, and in another lesson their behaviour might be quite different — it was the teacher causing the effect.

🖉 This answer gains full marks as two accurate reasons are identified and each is explained. In the second part of the answer, the example given in the last sentence does not logically follow on from the reason given — it still doesn't help us to understand why pupils might behave differently with one teacher than with another. However, there is already sufficient information for full marks to be awarded.

8/8 marks

(e) Questionnaires are usually used by sociologists in large-scale social surveys. They give mainly quantitative data (although open-ended questions give qualitative data) and so represent the positivist approach. They are useful because large numbers of people can be asked identical questions over a short period of time, and the answers to closed questions can be quickly and easily analysed.

🖉 This is a very good introduction which provides a lot of useful information in a short space — when questionnaires are likely to be used, which sociological approach they represent, the different types of data resulting from different styles of question, why questionnaires are useful, and something about the analysis of the data. The examiner can therefore see right from the start that this candidate has a good understanding of questionnaires. While this is a very useful kind of introduction, it is important to make sure that it is well focused and does not become too long.

Writing a questionnaire is not easy, and it is always a good idea to do a pilot study so that a small group of people can be asked the questions to see whether there are any problems before the questionnaire is given to all the people in the sample, which in some cases might run into thousands.

🖉 Another very useful point is made, further establishing the candidate's understanding of questionnaires.

One problem is that all the questions have to be clear, so that the respondents can understand them. It is not a good idea when using standardised questionnaires for the people to have to keep asking the interviewer what the questions mean, as if the interviewer starts to explain things this can cause bias. As well as having to use clear language, the questions should be ones that the people have the knowledge to answer; otherwise they might feel stupid and start to make things up, which would also cause bias.

🖉 Two good examples are given of the ways in which the questions should be clear to respondents, both in their wording and in their subject matter. The latter point is one that is often missed when discussing the possible problems of writing questionnaires.

Another thing to look out for is to make sure that the questions don't cause offence or embarrassment. Questions on sensitive topics such as domestic violence, sexual

abuse or bullying are usually better dealt with through interviews rather than questionnaires. Racist and sexist language should always be avoided. Questions should also be neutral — that is, the respondent is not pushed to answer one way rather than another. Questions starting 'Don't you think that...?' or 'Wouldn't you agree that...?' are pushing the respondent towards a particular answer.

✏ A further three important issues are raised, with a useful example to illustrate the third point.

You should also make sure that you are asking only one question at a time. For example, something like 'Do you like hospital dramas, and do you think that there are too many of them on television?' is called a 'portmanteau' question. A portmanteau is a large bag, and it means that the question is carrying more than one question within it. You would have to make this two separate questions and, in the second one, it would be better to give the person a choice, i.e. to choose between 'too many, just about right or not enough' hospital dramas.

✏ The candidate makes another well-illustrated point.

Finally, you should make sure that your questionnaire isn't too long. People are not usually paid for answering sociological questionnaires, and if the researcher is taking up too much of their time they might start to get fed up and not really think about their answers, as they just want to get away; or they might just stop answering the questions altogether, which would mean that the questionnaire couldn't be used. One way is to make sure that you only ask about the things that you really need to know about, or you will end up with a lot of information which you don't use, which is a waste of time and money.

✏ Another two important points are raised, one focused on the respondents and their possible reaction to an over-long questionnaire, and the other on the researcher possibly ending up with too much information.

So, as I have shown, while questionnaires are very important in sociological research, they are not easy to write and contain a lot of potential problems which sociologists have to try to avoid if they want the data to be useful to their research.

✏ This is a brief but appropriate conclusion that arises logically from the points that the candidate has made in the preceding paragraphs.

Overall, this is a well-focused and well-written answer that addresses a number of problems associated with writing good questionnaires. There are some other points the candidate could have raised — can you think of any other problems? What about dividing people into age groups, for example? Nevertheless, it is important to remember that in questions such as this, candidates are not expected to identify and discuss every possible thing related to the question — that is why the phrase 'some of the problems' is used. As long as a range of issues/ problems is raised, you will be able to get into the top mark band, assuming that your answer meets the other criteria of this band. There is a generally evaluative/critical approach running through this answer, especially in some of the examples given, so the candidate is demonstrating both assessment objectives. The answer meets the criteria of the top of the 15–20 band, and gets full marks.

20/20 marks

(f) The information in Item B is about a particular longitudinal study, the National Child Development Study. You can tell from its name that it is a very large-scale study, probably funded by the government. Longitudinal studies are studies which take a group of people and study them at fairly regular intervals over time. An example of a longitudinal study in sociology was Douglas's 'The Home and the School'.

🖉 This introduction shows that the candidate understands what a longitudinal study is, and can give an example of one from sociology. The candidate attempts to work out something about the study mentioned in the item. This shows that even if you are unfamiliar with the source of the material in the item, careful reading will always allow you to say something useful about it.

Longitudinal studies have some good advantages. A lot of sociological research is really just a 'snapshot', a picture of something at one particular point in time, but longitudinal studies allow the effects of something to be seen over time. For example, in Douglas's study, he was able to see what factors could have influenced the children's 11+ results, and could then go on to see how the different children got on in the different kinds of school they went to and the different streams they were put in. He was able to measure the effects of things in the children's home background, such as their family size and social class.

🖉 The candidate gives a clear illustration of one of the strengths of longitudinal studies, backed up by an example. Note that even if you are unfamiliar with this particular study, you should always be able to use an example of any particular method. Make sure that when you make notes on the different sociological methods, you include at least two examples of studies which have used that method, so that you can refer to them in exam answers.

As Item B says, in collecting information about the people in a group, you will end up with a mass of detail which can be used for a variety of things, such as medical/health research. In this study, the information collected on the people when they were children could be used to examine some reasons for differences in their health in later life — you could see whether children from poor homes were likely to have more illnesses than children from richer homes, or whether children in towns had more accidents than children in the countryside, for example. Collecting all the information in a systematic way would allow the researchers to see if they could pick out certain important variables which seemed to have a big effect on people's lives.

🖉 This is a sensitive and quite sophisticated attempt to interpret the material in the item to demonstrate an important advantage of longitudinal studies — the ability

to find important variables that exert significant effects. This candidate is really trying to work with the material in the item to answer the question.

However, longitudinal studies also have disadvantages. They are very expensive, especially when there are a lot of people in the group being studied. You need a team of researchers, not just one person. It would probably not be possible to keep the same team if the study took a long time, so you would have to be careful that the new people fitted into the team and carried on the research as it was planned. A huge amount of information is gathered and this would be very time-consuming to analyse, which also adds to the expense.

Here the candidate is not just identifying some disadvantages but is attempting to explain them. It is important to try to do this and not just give a list of advantages and disadvantages.

Another expense would be trying to keep track of all the people you were studying. As the item says, there have been five attempts to trace all the people in the group. Some people would drop out over time. They might move and not leave a new address, or might go and live abroad, or might just not bother to take part any more. This can affect the representativeness of the group, as the people who are 'lost' might be different from those that stay in the research. This is less of a problem where you have very large numbers to begin with, as in the example in Item B.

The candidate links another disadvantage to the important issue of 'representativeness', and makes another good link to the item.

Another problem is that for some things, the fact that the people were being studied over time could mean that they began to think about things differently, as they would know that they would be asked questions about them. This would not be a problem with things like their medical history or educational qualifications, but if they were asked about their political or religious views over time, they might come up with different answers because they were thinking about these things more. They might become different from people who were not in the study.

The candidate identifies another potential problem and again explains, with examples, why it might be problematic.

In conclusion, although longitudinal studies do have disadvantages, they are still very important, as they can tell us things that other types of study, that are a snapshot in time, can't. As they are so expensive, not many are carried out, which probably makes the ones that are carried out even more important to sociology.

This conclusion appears to come down on the side of longitudinal studies, implying that their advantages are sufficiently important to outweigh their disadvantages. While it is not always necessary, or even possible, to come to a distinct conclusion, if you consider that your arguments favour a particular view or method, don't be afraid to say so.

This is a good answer that remains focused on the question. The candidate might have mentioned that longitudinal studies can collect both quantitative and qualitative data, but not mentioning this does not have a serious effect on the marks gained. The answer shows extremely good use of the material in the item, which is closely applied to the question. Evaluation is shown in the attempt to weigh up the advantages and disadvantages of the method, and in the discussion of the reasons for these. **18/20 marks**

Overall: 58/60 marks

■ ■ ■

Answer to question 3: grade-C candidate

(a) The Hawthorne effect comes from the Hawthorne research in a factory and means that when a researcher is studying people they will change their behaviour and not act normally.

> ☑ Explaining the source of a concept is not necessary — what is asked for is an accurate definition. Unfortunately, the candidate fails to make clear the point that the Hawthorne effect occurs (or may occur) only when people are aware that they are being studied. **0/2 marks**

(b) Quantitative data can be displayed in tables and tick boxes.

> ☑ 2 marks are awarded for 'tables', but 'tick boxes', which the candidate has picked up from Item A, is not a precise enough answer to gain marks. **2/4 marks**

(c) It is sometimes hard to get permission to study the group. The researcher can become too familiar with the group and become biased. In covert PO it is not fair to deceive people and study them without their consent.

> ☑ The second and third reasons are awarded the marks. The first reason does not gain marks, as it is a possible problem for the researcher rather than a criticism of the *method*. However, it could have been expressed in a way that would have been an appropriate criticism of the method — can you think of how this might be done? (Clue: it is about the representativeness of the groups studied.) Note that the candidate is careful to make it clear that the third point refers specifically to *covert* participant observation. **4/6 marks**

(d) If you are trying to look at a whole classroom to see who is doing what, if it is a large class then you can miss some things, as you won't be able to see all the pupils all of the time. If you don't know the school or the teacher, you could put the wrong interpretation on things. For example, you might see that the teacher kept asking questions of particular pupils, say one or two boys, and you might think that this was gender bias against girls, but if you spoke to the teacher you might find out that she was trying to encourage some shy or timid boys to speak up to show them and the others that they did really know the answers, so it wouldn't be a wrong thing (bias) but a good thing (trying to help particular pupils).

💬 It would have been better if this answer had been separated into the two component parts. However, two appropriate reasons are identified and each is further explained, so full marks are achieved. **8/8 marks**

(e) There are several problems to do with writing a good questionnaire, and sociologists have to bear these in mind, otherwise their research will not be as good as it should be.

💬 This is not too bad for a brief introduction, although it would have been much better if the candidate had explained in more detail what was meant by the research not being 'as good as it should be'.

Questionnaires are used in social surveys, and can be given by an interviewer or sent through the post. Postal questionnaires cause problems as they have a low response rate, and you can never be sure who has filled them in — people could have given them to another member of the family, or even a next-door neighbour, and the sociologist won't know this.

💬 While the first point is potentially relevant, the candidate then gets side-tracked into some of the problems of postal surveys, rather than keeping the focus on the construction of questionnaires. Make sure that you always keep in mind what the question is about, or what you write, even if accurate, will not score you marks as it will not be relevant to the question set. There is a point about postal questionnaires that is relevant to the question — can you think what this might be?

One of the problems with writing questionnaires is that you should make sure that the questions are clear and that the people can understand what you are asking them. This means you should avoid long words and technical terms. Another problem is that you have to avoid offensive or embarrassing questions, like 'Do you think you need a deodorant?' or 'Do you think you are too fat?'

💬 Two appropriate problems are identified, with some explanation. It is a pity that the candidate didn't give examples in the second part of the paragraph that were more 'sociological' in nature.

You should also realise that some people don't like being asked about their age, especially women, and you should avoid questions that are sexist or racist. Questionnaires also shouldn't be too long, or people will get bored with them and not treat them seriously.

💬 Two further points are made, each of which could have been developed, particularly the point about 'sexist or racist' questions. So far, the answer is still at a rather simplistic level, and there is not much evidence that the candidate has been on a sociology course.

Some researchers (positivists) like questionnaires because they think they are more 'scientific' because they are collecting the facts. This is true up to a point, especially if you are using closed questions. Other researchers (interpretivists) disagree, and say that questionnaires don't give people the chance to express their feelings or say why they do things. It is also claimed that questionnaires are reliable, so that if you did the survey again you would get the same results, but this might depend on what you were asking about and who you were asking.

The answer begins to become more sociological in nature, and makes some potentially relevant and important points, but again these are not developed. The very last point gives a hint of evaluation, but needs explaining.

Questionnaires can get a lot of information from a lot of people in a short space of time, so they are useful in sociological research, and many sociologists will continue to use them. The Census is an important questionnaire given out by the government to every household in the country, and everybody has to fill it in by law.

The first point could have been developed further, and the example of the Census, which is potentially relevant, is lost because there is no focus on the construction of questionnaires.

The candidate shows a reasonable knowledge and understanding of some potentially relevant material, which puts it in the 8–14 mark band for (e) questions. However, many of the points made are fairly simplistic and there is little wider discussion. The answer is not supported by examples from sociological research and fits in the lower, rather than the higher, part of the mark band. **9/20 marks**

(f) Longitudinal studies, like the one in Item B, take a group of people and follow them over a period of time. The people in the National Child Development Study will now be in their forties, so it is amazing to think that they have been studied all this time.

Reference is made to the item, and the candidate uses the information to make a point about the length of time this study has gone on.

There are some advantages to this kind of study. The researchers will know quite a lot about the people in the study, as they have been studying them from birth, so when they do their research they will not have to keep going back and asking them the same questions as they will know the answers already. This is a bit like the *Seven Up* programmes on television, where a group of children have been studied since they were seven, and in the last programme they were 42. Of course, there aren't many people in the *Seven Up* programmes, but in Item B it says that there are 17,000 people, so the researchers won't know these as well as the television children.

A relevant point is made about knowing the people in the survey. It will be interesting to see whether the candidate acknowledges that this might be a disadvantage to longitudinal studies, as well as an advantage. An interesting comparison is made with a different type of longitudinal study — one whose prime aim is entertainment.

The *Seven Up* study is looking mainly at social class, as the children come from different backgrounds. The Item B study would be able to do the same, as they would know what backgrounds the people in their survey came from. They could see what kind of education the different people had, and what qualifications they got, and whether they stayed in the same social class.

Again, the candidate makes comparisons between the two studies, using social class and showing how the effects of this on education might be examined.

Longitudinal studies also have disadvantages. Some of the people might not want to carry on taking part, like some of the *Seven Up* children, who aren't in the series any more. They cost a lot of money, especially if there are a lot of people to be interviewed, and it must have taken a lot of time to contact all the schools and colleges the people in the Item B study went to. Also, people might start to behave differently, for example one of the *Seven Up* people says that now they start to think more about what is happening to them because they know that it will be shown on television at some point. This would be one difference, as they have come to be known by millions of people, whereas the people in the Item B study are probably kept anonymous when the research findings are written up and published.

Three disadvantages are mentioned, although the first one could have been expanded into a discussion of the implications of some people 'dropping out' of the original group. Although the candidate is using just one other example of a longitudinal study, an attempt is made to relate the points to the study described in the item. An interesting final point is made, showing that the candidate is trying hard to interpret aspects of the two studies described in the answer.

Therefore, longitudinal studies have advantages and disadvantages, just like other methods. As it says in Item B, if it is a big study, the information can be used for different things, like medical and health research, which could be very important.

This is a brief and not particularly helpful conclusion, which adds little to what has been said before.

A reasonable attempt to answer the question has been made, showing some limited analysis and evaluation. The candidate tries hard to make use of the information in the item and is able to make points of comparison with another study. Relatively few points are made about the advantages and disadvantages of longitudinal studies, and the points that are made could have been further developed.   **10/20 marks**

Overall: 33/60 marks

AS
Coursework Task

This is a structured task that asks you to submit a proposal for a piece of sociological research. The proposed research does not have to be something that could be carried out by a 17-year-old school or college student. However, if you think that you may wish to use your AS proposal as the basis of an A2 piece of coursework, your AS work will obviously need to be something that you could realistically achieve as a research project in the second year of your course.

The proposal has to be submitted under four separate headings, and each section has its own word length, with the total number of words not exceeding 1,200. It is important to remember that, although each of the four parts has its own mark scheme and deals with a different aspect of the research, the proposal is read by the examiner as a whole piece. In practice, this means that each part of your proposal follows logically on from the previous one. As a result of this, the most important thing to get right is your initial hypothesis or aim, as the examiner will use this to see how it can be 'tracked' through the rest of the sections. To spell this out even more clearly, you should be able to state:

- some sociological reasons why you chose your particular hypothesis or aim
- how your context pieces are linked to it
- how your chosen concepts are ones that you would need to use in the design, carrying out and analysis of the research
- how and why the method chosen is, in your opinion, the most suitable for investigating that particular hypothesis or aim
- how the 'potential problems' are linked not just to the chosen method as such, but to the method as employed in investigating your chosen research focus

In this section of the unit guide, each of the four coursework task headings is taken separately, and advice given on how to tackle that part of the task. In addition, an example is given of a candidate's initial ideas about a particular coursework task. This is not a draft of a coursework task, but the first rough proposals to be discussed with a teacher. It is accompanied by comments and suggested tasks for you to do. You should read each subsection in turn and attempt the tasks, and then apply what you have learned to your own research proposal. At the end of each subsection, the mark-band descriptor is given for the top band in that part of the task. Read this carefully, to help judge whether what the candidate has written is likely to meet the stated criteria. This will also make you aware of what it is that you will have to achieve to gain top-band marks.

Preparing for the coursework task

Title

This is not part of the coursework task, and is not awarded any marks, but most students feel more comfortable if their work has a title. It is probably better to start with a 'working title', which can be modified later if necessary, than to spend a long time trying to think of a witty or very academic-sounding title. Too much detail in the initial title can lead to a distortion of the task, as it will push you in a particular direction, encouraging you to match the proposal to the title, rather than the other way round.

The piece of candidate's work to be considered throughout the following sections has the working title:

How parents choose a secondary school for their children

It is quite possible that the candidate will not use this as the finished title, but it is sufficient for the moment, in that it shows clearly the focus of the proposed research.

Hypothesis/aim

Candidate's proposal

I think that I will choose an aim rather than a hypothesis, as I am not sure what I will find. I want to look at something about how parents go about deciding on a secondary school for their child. I have enjoyed my work on the sociology of education, and think that I will find developing my aim very interesting. Parents seem very important to the success or otherwise of their child at school. Social class was (is?) a major factor here, with middle-class parents usually having higher aspirations than working-class parents, and generally knowing much more about how the system operates. Is this still the case? Now that league tables are available for all parents to look at, perhaps working-class parents have become more aware of the differences between schools. A good moment to explore this is as the children are about to move to secondary school.

> As stated above, this is not what the candidate intends to present for this section, but rather the initial ideas. A reason is given for deciding to have an aim rather than a hypothesis, and the candidate is able to present some good supporting material for choosing an aim exploring factors affecting parental choice. Given the tight word count, when the work is finally written up, the

second sentence should probably be left out, as it is vague and not particularly helpful. The third and fourth sentences are much more relevant to the choice of the aim.

Top-band descriptor

'The candidate demonstrates a very good ability to identify and define a relevant sociological focus. A clear, precise and appropriate hypothesis or aim that enables the research issue to be successfully progressed is offered. All reasons given are appropriate.'

Tasks

1 The candidate hasn't actually presented a specific aim. Write an aim which is suitable for a sociology research proposal. The candidate's choice of an aim rather than a hypothesis is quite appropriate in this case. However, try to think of a hypothesis that could have been offered instead.

2 Without looking at the next subsection, jot down a couple of pieces of material that could be used as context pieces for the research proposal. Remember that the chosen aim or hypothesis must be able to be 'tracked' through all the subsequent parts of the proposal, so these should all relate back to the aim/hypothesis.

3 Again without looking at the next subsection, jot down two appropriate concepts that might be used in the research proposal.

Context and concepts

Candidate's proposal

Most schools are now comprehensive, so the grammar/secondary modern school divide has gone. Allocation to comprehensive schools used to be by catchment area, but recently it has been largely on the basis of parental choice. How do parents choose? My contexts are:

1 J. W. B. Douglas, *The Home and the School* (1964). Douglas was writing when there was the tripartite system. He found that over half the mothers in his survey wanted their children to go to grammar school, even though under 20% of all children got grammar school places. He also found that mothers' aspirations seemed to have an effect when grammar school places were awarded, especially if the children were 'borderline' in terms of ability.

2 Two related articles in the *Guardian* (17 September 2000 and 2 August 2001) expressing concern over the effects of parental choice. In the first article, teachers claimed that the 'yuppie Land Rover brigade' are ferrying their children miles to the 'best' state schools, leaving other schools to become 'sink' schools. The second article is about a report from the Institute for Public Policy Research, which

claims that parents from all socioeconomic backgrounds are appealing when their children don't get into the school of first choice. The report says that parental choice is 'generally beneficial', but can have a 'significant negative impact' on many schools, as those lower down the school league tables suffer from falling intake and get fewer resources. Disaffected and difficult children become concentrated in these schools, creating a 'vicious circle'.

> The context pieces are both focused on the chosen area, i.e. parental choice of school, though they describe different educational systems. While the 'second' context piece actually refers to two newspaper articles, these are obviously related. It can be more difficult to find a single article that provides the necessary context, but you should try not to provide more than two, and these should, as here, be closely related. There is a mention of 'league tables' in one of the newspaper articles, which is probably linked to the notion of the 'best' state schools mentioned in the other. Although not a requirement, it is usually a good idea to have at least one of your pieces of context material based on a 'sociological' text, particularly with a topic such as this.

My concepts are 'parental aspirations' and 'social class'. I have chosen 'parental aspirations' because I want to explore whether parents choose a particular school because they think that their child will receive a better education there and do well in education. Douglas's research showed that there was a clear link between the level of parental aspiration and a child's performance, even allowing for social class. I have chosen 'social class' because I want to see whether middle-class parents know more about the league tables than working-class parents, and also whether they have higher aspirations for their children.

> The candidate has chosen two appropriate concepts, but remember that these will have to be 'carried through' into the next parts of the proposal. Both concepts are appropriate to the initial aim.

Top-band descriptor

'The candidate demonstrates a very good knowledge and understanding of appropriate material. Two relevant pieces of material are accurately and concisely presented, providing a clear context for the proposed research hypothesis or aim. The two or three concepts identified are pertinent, precisely defined and appropriately developed.'

Tasks

1 Carry out an internet search to find other material from newspapers which could also have been used as context pieces.

2 Explain what is meant by 'parental aspirations'.

3 Identify a potential problem with choosing 'social class' as a concept.

4 Compare the candidate's choice of concepts with your own. If the concepts were different, think carefully about which ones are likely to be better. What reasons would you give for this?

5 Without looking at the next subsection, jot down which research method you would choose to research the candidate's aim, with brief reasons for your choice.

Main method and reasons

Candidate's proposal

I would use standardised questionnaires as my method. These would contain both closed and open-ended questions. I would need to find out whether a school had already been chosen (by the time the child was about 8 years old) and why the parent had chosen it (could be for more than one reason). This should help me to identify the level of parental aspiration. I would have to find out whether they were aware of the chosen school's position in the league table. I would ask how far away the chosen school was, and whether there were schools nearer to the parent's home, and why these had not been chosen. I would probably include questions to find out how important they thought education was. I would also ask them to rank certain characteristics in order of importance to them, e.g. discipline, single-sex or mixed, school uniform, extra-curricular activities, school resources and facilities, etc.

In order to avoid asking direct questions about social class (i.e. about occupation), I would do my questionnaires by choosing four different primary schools and standing at the gates when the mothers came to collect their children (as the children would be younger, they would probably still be met by a parent to be taken home). I would choose two primary schools in working-class areas and two in middle-class areas. I would try to get 25 completed questionnaires from each school, i.e. 100 in all.

Reasons for choice of method: this research would mainly be a first gathering of information, with the results possibly being used to develop and test a hypothesis later. It would be more helpful, therefore, to collect a fairly large amount of mainly quantitative information than a smaller amount of in-depth qualitative information.

> The candidate has provided some details about, and reasons for, the choice of method. Reference has been made to the chosen concepts. Although a clear aim was not stated in the first part of the proposal, the material here clearly relates to an exploration of the reasons for parents choosing a particular secondary school.

Top-band descriptor

'The candidate demonstrates a very good knowledge and understanding of the chosen method. All reasons for the choice of method are relevant and developed. All details on the implementation of the proposed methodology are appropriate, clear, accurate and succinctly presented. The chosen method is appropriate to the hypothesis or aim.'

1 Identify some further details that the candidate might need to provide (bearing in mind the word limit) to show 'a very good knowledge and understanding of the chosen method'.

2 Without looking at the following subsection, identify some potential problems with the method as outlined by the candidate.

3 Using the information provided by the candidate, draft a questionnaire that could be administered to parents. (Note: while you are not asked to provide an example of a questionnaire or interview schedule, and should not send one with your coursework task, it is always a good idea to draft one if this is your chosen method. This will help you to see whether the information you have provided in this part of the proposal is suitable and covers the topics you would want to find out about. It will also help you to see whether some of your points are vague or unrealistic, e.g. 'how important they thought that education was'. Moreover, it will prove useful practice if you intend to use one of these methods in A2 coursework.)

Potential problems

Candidate's proposal

- I won't be able to be very sure about the 'social class' categories.
- It would take a long time to ask the questions and then analyse the results from 100 questionnaires.
- Some people could be too busy to answer my questions.
- As with all questionnaires, people might not tell me the truth.
- It would be difficult for other people not to overhear the answers being given.

> The candidate has listed some potential problems, but none of the points has been developed. In other words, there is no real evidence that the candidate understands clearly why each of these is a potential problem. Although there is a fairly tight word limit, you should always make it clear why something might be a problem — again, always in terms of your initial aim or hypothesis.

Top-band descriptor

'The candidate demonstrates a very good ability to identify potential problems in carrying out the proposed research. Appropriate, accurate and succinct reasons are offered and explained. There are clear links between the identified problems and the research hypothesis or aim. The candidate demonstrates very good sociological insight.'

1 The candidate has made no reference to 'sampling'. If you have not already done so, identify the point that should have been made about this.

2 If this proposal were to be developed into a piece of actual research at A2, what point would you make about the number of questionnaires to be administered?

Go through the bullet points above (and any others you have identified) and briefly explain why each of them might be a problem.

3 Go back to the original aim (or your interpretation of it) and check each part of the proposal to see whether the aim can be 'tracked' through it. Think carefully about what information the answers to the questionnaire might provide — would this enable the aim to be fulfilled? If not, there are two possibilities. One is to modify the aim so that the information gathered is appropriate. The other is to keep the initial aim and look again at the data-collection process identified, to see whether any part of it should be modified.

Although clearly not in finished form, this has the potential to be a good research proposal. However, the candidate would need to do quite a bit of extra work to ensure gaining high marks.